T0188264

Physics and Chemistry
of Nanostructured
Materials

Physics and Chemistry of Nanostructured Materials

Edited by

Shihe Yang and Ping Sheng

Hong Kong University of Science and Technology
Clear Water Bay, Hong Kong

CRC Press
Taylor & Francis Group
Boca Raton London New York

CRC Press is an imprint of the
Taylor & Francis Group, an **informa** business

A TAYLOR & FRANCIS BOOK

CRC Press
Taylor & Francis Group
6000 Broken Sound Parkway NW, Suite 300
Boca Raton, FL 33487-2742

First issued in paperback 2019

© 2000 Shihe Yang and Ping Sheng
CRC Press is an imprint of Taylor & Francis Group, an Informa business

No claim to original U.S. Government works

ISBN-13: 978-0-367-39928-3

This book contains information obtained from authentic and highly regarded sources. Reasonable efforts have been made to publish reliable data and information, but the author and publisher cannot assume responsibility for the validity of all materials or the consequences of their use. The authors and publishers have attempted to trace the copyright holders of all material reproduced in this publication and apologize to copyright holders if permission to publish in this form has not been obtained. If any copyright material has not been acknowledged please write and let us know so we may rectify in any future reprint.

Except as permitted under U.S. Copyright Law, no part of this book may be reprinted, reproduced, transmitted, or utilized in any form by any electronic, mechanical, or other means, now known or hereafter invented, including photocopying, microfilming, and recording, or in any information storage or retrieval system, without written permission from the publishers.

For permission to photocopy or use material electronically from this work, please access www.copyright.com (http://www.copyright.com/) or contact the Copyright Clearance Center, Inc. (CCC), 222 Rosewood Drive, Danvers, MA 01923, 978-750-8400. CCC is a not-for-profit organization that provides licenses and registration for a variety of users. For organizations that have been granted a photocopy license by the CCC, a separate system of payment has been arranged.

Trademark Notice: Product or corporate names may be trademarks or registered trademarks, and are used only for identification and explanation without intent to infringe.

British Library Cataloguing in Publication Data
A catalogue record for this book is available from the British Library

Library of Congress Cataloging in Publication Data available upon request

Visit the Taylor & Francis Web site at
http://www.taylorandfrancis.com

and the CRC Press Web site at
http://www.crcpress.com

Contents

Part 3   FUNCTIONAL NANOSTRUCTURED MATERIALS

Part 4 NANOCOMPOSITES

Preface

Evolution of human civilization, from the Stone Age to the age of silicon, is inseparable from materials development. Today, more than ever, innovation of new materials represents one of the most important driving forces for new technologies and economic progress. Against this background, the emergence of nanostructured materials is especially significant, as the synthesis of materials with nanometer-scale entities has created a wide range of novel properties unattainable in bulk materials, with technological implications yet to be fully explored. The unifying concept in nanostructured materials is that novel materials properties may be induced through the formation/fabrication of nano- to micro-scaled structures. Under this simple concept, however, lies a diversity of possibilities introduced through parameters such as the spatial dimensionality of the structures (e.g., two-dimensional layers, one-dimensional wires, or zero-dimensional clusters), the size and the atomic composition. Fueled by this rich brew of potential structures/properties, the recent explosion of global activities in nanostructured materials may be compared to the initial stage of an evolutionary process, in which the novel materials are continuously being evolved/assembled into higher-level structures with new capabilities, and in which the global market place acts as the ultimate arbiter for survival. In this analogy, it would perhaps be futile to speculate on the specific technological or social impact implied by nanostructured materials, but there is no doubt on their increasingly important role in our future.

The organisation of this Advanced Study Institute (ASI) on the Physics and Chemistry of Nanostructured Materials grew out of discussions among the organizers in late 1997, prompted by an invitation from the Croucher Foundation to submit a proposal. The organization of this ASI served not only to review and capture a snapshot of this important field, but also to underscore Hong Kong's unique role as a window for scientific exchange, as well as its new commitment to research and development. It is the organizers' hope that this ASI and its proceedings will serve as a medium for cross-pollination, so that even more scientific fruits may result in the future.

This proceeding collects both the invited, oral and poster contributions to the Croucher ASI workshop. It covers preparation and characterization of nanostructured materials as well as the relevant physics and chemistry. The text is organized into four sections: i) metal and semiconductor nanostructures; ii) carbon nanostructures; iii) functional nanostructured materials; and iv) nanocomposites.

We wish to acknowledge the Croucher Foundation for its generous financial support. The successful organization of this ASI owes much to the efforts of Louiza Law, Helen Lai, and many others. To them we wish to express our heartfelt thanks.

Shihe Yang and Ping Sheng
Clear Water Bay, Hong Kong
March 1999

Part 1

METAL AND SEMICONDUCTOR NANOSTRUCTURES

1   Semiconductor Nanocrystal Colloids

A. Paul Alivisatos
*Department of Chemistry, University of California, Berkeley, USA and
Lawrence Berkeley National Laboratory*

1.1 SEMICONDUCTOR COLLOIDS

The study of nanometer size semiconductor crystals has been advancing at a rapid pace[1,2]. Much of the interest in these materials stems form the fact that their physical and chemical properties can be systematically tuned by variation of the size, according to increasingly well established scaling laws. This paper describes colloidal semiconductor nanocrystals belonging to the II-VI and III-V families, and outlines strategies for obtaining electrical access to such dots.

If an inorganic cluster exceeds a certain size, generally in the tens of unit cells, then it will likely possess a bonding geometry characteristic of a bulk phase[3,4]. Above this critical size, the nature of the chemical bonds in the cluster remains fixed as a function of the size, but the total number of atoms, or the surface to volume ratio, change smoothly. This leads to a slow extrapolation of the properties of ideal nanocrystals towards bulk values with increasing size, according to the scaling laws. The ability to systematically control the properties of inorganic materials by variation of size and shape is an important development, with many implications for how materials should be processed and assembled[5].

Many scaling laws have been investigated, including the size variation of band gap, charging energy, magnetization reversal, melting, etc. Study of the scaling laws reveal lessons for how to make nanocrystals. This paper focuses on the properties of colloidal semiconductor nanocrystals, how to make them, and some ways of gaining electrical access to them. Advances in metal[6-8], magnetic[9], and structural[10-12] nano-materials are also occurring. Semiconductor dots produced by other processing techniques, such as self assembled semiconductor dots, and porous silicon, are reviewed elsewhere in this issue.

1.2 PROCESSING ON THE NANOSCALE

Many routes have been developed for preparing nanostructures, ranging from single atom manipulation to organic synthesis. The tremendous strengths of the colloidal preparation routes are that they yield large amounts of monodisperse nanocrystals with easy size tuning. Further, the presence of the organic molecules on the surface of a nanocrystal enable extensive chemical manipulation after the synthesis. One interesting conclusion from this is that the preparation of nanocrystals may in many ways be far more forgiving than the fabrication of two or three dimensional materials.

1.1.1 Nanocrystals can be prepared at comparatively modest temperatures.

One of the most famous scaling laws relates the variation of melting temperature with size, and shows that it drops as $1/r$[13,14]. The large drop in melting temperature in small sizes, means that it is possible to make highly crystalline, and faceted nanoparticles at temperatures that are compatible with wet chemical processing. From a kinetic perspective, a great deal of control of chemical processes can be achieved in solution, and for this reason extremely high quality inorganic nanoparticles can be prepared as colloids. The recent successes in the preparation of II-VI[15,16] and III-V[17-20] nanocrystals illustrate the strengths of the colloidal preparation techniques.

Nanocrystal precursor molecules are injected into a hot surfactant at 300C;

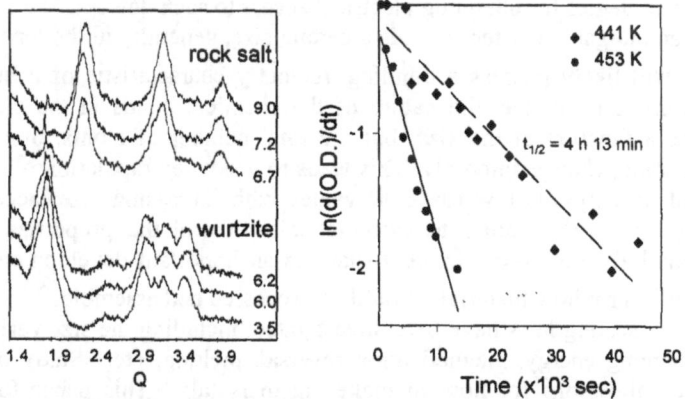

Figure 1. A. X-ray powder diffraction patterns from 4 nm CdSe nanocrystals under hydrostatic pressure shows that they convert from four coordinate wurtzite to six coordinate rocksalt at pressures far higher than the bulk (6. 5 vs. 2.7 GPa) (Data taken as part of the UC-National Lab PRT beamline at the Stanford Synchrotron Radiation Lab.) B. When the pressure is increased abruptly, the phase transition proceeds in a characteristic relaxation time. The temperature dependence of this relaxation time yields the barrier height blocking the transition

nucleation and growth of the nanoparticles occurs. There is no aggregation, because the particles are terminated by a monolayer of surfactant. At the end of the preparation, nanocrystals are isolated with a monolayer surfactant coat, but this can be exchanged with other organic molecules, allowing the environment of the nanoparticles to be controlled.

1.1.2 Nanocrystals act as single structural domains.

Structurally, nanocrystals are at a crossover point between molecules, where the number and location of each atom can be exactly specified, and bulk matter, where we must work with statistical descriptions. Both kinetically and thermodynamically, nanocrystals tend to exclude high energy defects, such as grain boundaries (lower energy defects, such as stacking faults are common, however). It is far easier to anneal out a defect from a very small crystal than from a large one. Assuming the barrier to defect migration is the same as the bulk solid (certainly an overestimate, since the melting temperature is lower), then the defect simply has less distance to travel and will be annealed out faster in a smaller crystal. Thermodynamically, defects are present in a large crystal of atoms at finite temperature . However, if the free energy of defect formation is relatively constant with size, then a very small solid is less likely to contain equilibrium defects.

The high quality of colloidal dots is confirmed by experiments on the kinetics of pressure induced solid-solid phase transitions in nanocrystals[21,22] (Figure 1). In extended solids, pressure induced solid-solid phase transitions nucleate at high energy defects. In nanocrystals such phase transitions are kinetically suppressed because of the absence of defects. Under excess pressure beyond the thermodynamic transition point, nanocrystals are observed to undergo structural transformations on a much longer time scale than the corresponding bulk crystal. A simple scaling law, consistent with recent experiments on 2 to 5 nm dots, shows that this relaxation time for structural transformation is proportional to nanocrystal size. This is analogous to the well known scaling law for magnetization reversal time scales in single domain magnets. Larger crystals (>10 or 20 nm) inevitably contain defects, which act as nucleation sites for phase transitions, much as domain walls lower the barrier for magnetization reversal. An important conclusion from these studies of structural stability and phase transitions versus size, is that very small, nanometer size crystals may in many ways be more perfect than extended, or two dimensional solids, even though the processing is less complex[*].

---

[*]It is of course possible to make defective or aggregated nanoparticles. Simple scaling laws, however, suggest that the relatively modest processing conditions of colloidal routes are sufficient to prepare extremely high quality inorganic nanocrystals.)

1.3 QUANTUM SIZE EFFECTS IN ELECTRONIC SPECTRA

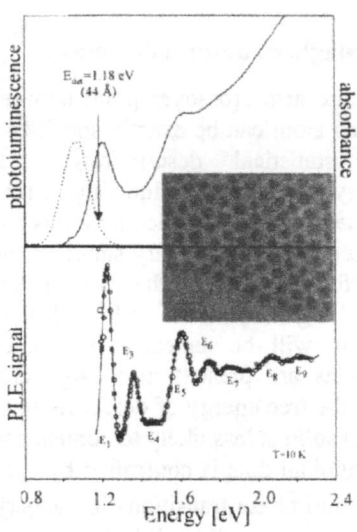

Figure 2 Optical spectra and transmission electron micrograph of InAs nanocrystals. Due to quantum size effects, the absorption and emission spectra are shifted to higher energy than in the bulk, and photoluminescence excitation spectra reveal the presence of multiple discrete transitions. The TEM shows that nanocrystals are monodisperse, and do not aggregate because of the monolayer surfactant coat on the surface.

The most widely investigated group of scaling laws in semiconductor nanocrystals concern the size evolution of the optical spectra. Many studies have focused on the properties of colloidal dots[23], or dots embedded in glass matrices[24]. Those studies have centered on II-VI dots of CdS and CdSe. A parallel set of studies have been directed towards the investigation of MBE grown dots, the majority of these studies being on III-V nanocrystals, such as InAs embedded in GaAs (reviewed in this issue). With the advent of colloidal synthetic routes to InAs and other III-V particles, these two parallel strains of research are now converging, and offer a fascinating opportunity to study nanostructured materials prepared by entirely different processes. Figure 2 shows the optical spectra of 5 nm diameter colloidal InAs dots. The electronic absorption spectrum is shifted to much higher energy than the bulk band gap, and the luminescence excitation reveals discrete structure in the spectra.. The systematic investigation of the evolution of the electronic structure with size remains an important topic of research, and a great deal of theoretical effort is directed at understanding this size evolution (see the article by Zunger in this issue).

1.4 CORE-SHELL DOTS

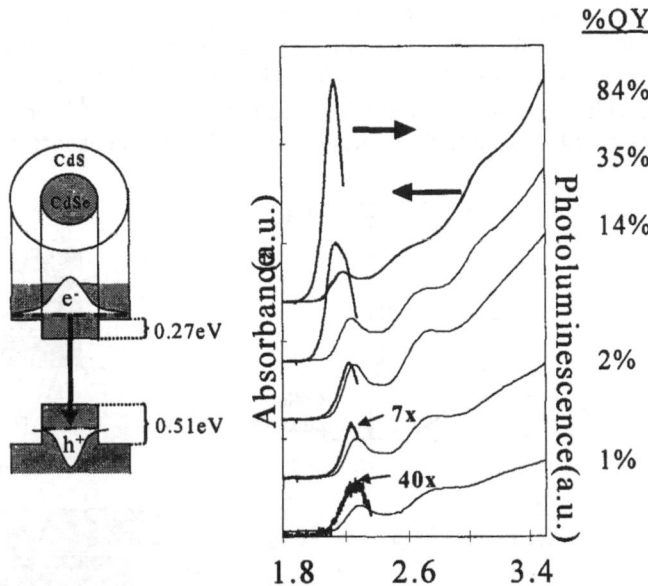

Figure 3 Core-shell nanocrystals of CdSe/CdS show greatly enhanced luminescence quantum yields. The outer shell of, CdS, is epitaxially deposited on the core of the CdSe.

One important issue for semiconductor nanocrystals concerns the nature of surface passivation. It is essential in a confined semiconductor structure to epitaxially passivate dangling bonds at the surface. Epitaxy and passivation in colloidal dots have now been achieved, making these dots fully comparable to MBE grown structures in surface quality[25]. As an example, CdS has been used to produce a fully epitaxial shell around CdSe[26]. That system displays band edge luminescence quantum yields as high as 80% at room temperature. (Figure 3). The outer CdS shell is still coated with a surfactant layer, so that the advantages of passivation and chemical processing are simultaneously realized.

In comparison to 2D structures, the kinetic energies in II-VI or III-V quantum dots of about 5nm diameter are of order 0.3 eV, so that the potential steps for electrons and holes at the hetero-interface need to be relatively large (order 1 eV) to be effective in confining the charges. Suitable passivation will be realized with different, and more dissimilar, pairs of materials in the 0D case, as compared to the 2D. Free standing dots also appear to be able to relax strain more readily than 2D structures. or embedded dots.

1.5 NANOCRYSTAL-POLYMER COMPOSITES

The electrical characteristics of quantum dots are of great interest, and we are pursuing two broad strategies for investigating them: studies of ensemble electrical characteristics, and electrical studies of individual dots. The most straightforward way to access an ensemble of colloidal nanocrystals electrically is by making a blend with a conducting or semiconducting polymer[27-29], onto which electrodes may be readily evaporated. As test cases, we have used nanocrystal/polymer blends

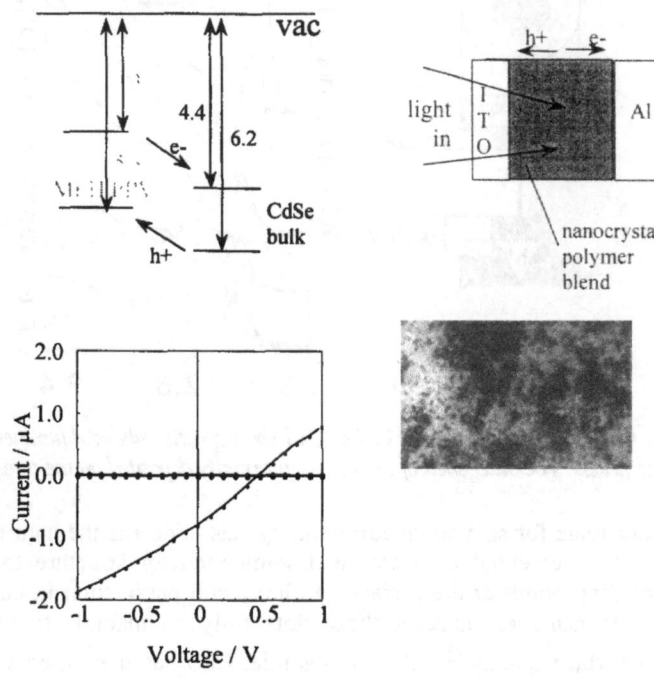

Figure 4 Schematic illustration of a photovoltaic based on an interpenetrating network of 5 nm diameter CdSe nanocrystals and the semiconductor polymer, MEH-PPV.

to make   photovoltaics and light emitting diodes.

1.6 PHOTOVOLTAICS

Nanocrystals are well suited to charge separation devices, since they provide a high surface area. The ability to tune the band gap readily by size variation, suggests that inexpensive multi band-gap solar cells, analogous to very high efficiency but very expensive quantum well based photovoltaics,   may be realizable with nanocrystals. O'Regan and Graetzel[30]   have reported an efficient (12%) and inexpensive photovoltaic based on a film of partially fused nanocrystals of TiO_2 coated with an extremely robust sensitizer molecule. That device takes advantage of the large surface area the nanoparticles provide, but does not make use of the quantum size effect to tune band gaps.

We have recently reported a photovoltaic device based on 5 nm CdSe nanocrystals and a semiconductor polymer, MEH-PPV [31](Figure 4). Both materials are semiconductors with band gaps of about 2eV, but the electron affinity of the nanocrystals is much greater than that of the polymer. As a consequence, electron-hole pairs generated with photons, either in the nanoparticles or the polymer, rapidly separate, with the hole residing on the polymer, and the electron on the nanoparticle. The nanocrystal/polymer blend morphology can be adjusted so that the nanoparticles form aggregated chains inside the polymer. This creates a pathway for holes to move along the polymer phase, and for electrons to hop (slowly) from nanocrystal to nanocrystal, until the separated charges are collected at electrodes. The efficiencies of these devices are quite low, and the fundamental physics of charge hopping and conduction in nanocrystals systems is still being studied, but it is clearly feasible to make interesting quantum dot based charge separation devices.

1.7 LEDS

We have also reported light emitting diodes based on nanocrystals and polymers[32-34]. Rather than making a blend, in this case we used a layered system, with a film of nanocrystals spin cast from solution onto a layer of semiconductor polymer, PPV. In this configuration holes are injected into the polymer, and electrons into the nanocrystals. Electron-hole pairs recombine by the emission of a photon in a zone near the nanocrystal-polymer interface. If the nanocrystal layer is relatively thick, then essentially all of the light comes form the nanocrystal layer only. In LEDs with moderate thickness nanocrystal layers, the recombination zone shifts with applied bias between the polymer and the nanocrystal layer, providing a form of voltage dependent color. Using a Mg/Ag electrode, we have achieved internal quantum efficiencies of 1%, and lifetimes in excess of several days. These LEDs combine the advantages of organic and inorganic LEDs. They are fabricated inexpensively in a way that can be scaled, and on flexible substrates, yet the active component, the nanocrystals, are very robust compared to most organic chromophores.

1.8 SINGLE NANOCRYSTAL SINGLE ELECTRON TRANSISTOR

Preliminary studies of both the electrical[35] and optical[36-39] properties of individual nanocrystals have been performed. These studies show that a single excess charge on a nanocrystal can dramatically influence its properties. We have recently reported measurements where the charge state of a single nanocrystal can be directly tuned[40]. As studies of lithographically patterned quantum dots and small metallic grains have shown, such measurements are invaluable in understanding the energy level spectra of small electronic systems. Back of the envelope estimates show that the charging energy of a 5.5 nm CdSe dot should be ca. 0.1 eV, and increases as $1/r$. The spacing of the discrete electronic levels of the dots scales as $1/r^2$, and coincidentally is also order 0.1 eV according to simple estimates.

Figure 5A shows an idealized schematic of the device. CdSe nanocrystals of a 5.5 nm diameter are bound to two closely spaced Au leads using bifunctional linker molecules. The leads are fabricated on a degenerately doped silicon wafer which is then used as a gate to tune the chemical potential, or charge state, of the nanocrystal

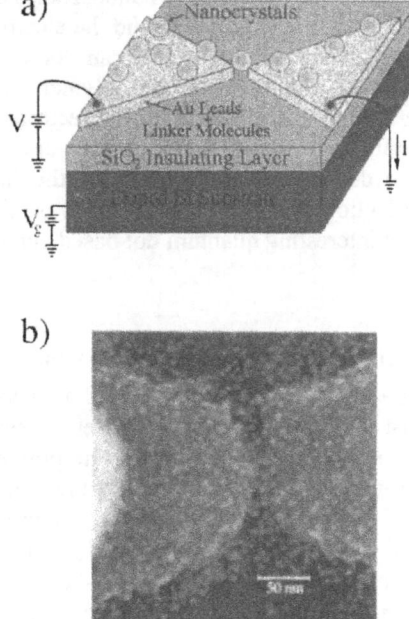

Figure 5.   A.   Schematic illustration of a transistor with one nanocrystal of CdSe as its active element. B.   Scanning electron micrograph of such a device.

under study.   Figure 5B shows a field emission scanning electron micrograph of a completed device. A number of nanocrystals appear to be in the ~ 5nm gap between the electrodes.

Most devices, including the particular junction shown in figure 5B, have immeasurably high impedance (R > 100 GΩ). Only about one in twenty have a measurable conductance, typically in the 10 MΩ -1 GΩ range. Devices of this type typically behave as though transport is occurring through a single nanocrystal. This may initially seem surprising, since Figure 5B indicates that the number of nanocrystals in the junction region is quite large. However, tunneling through the linker molecules has an exponential decay length of less than 1 Å. As a result, only a well-placed nanocrystal (within 2 nm of each lead) can contribute to conduction.

The current-voltage characteristics of these devices   at 4.2 K show characteristic single electron transistor behavior (Figure 6). At some arbitrary value of the gate voltage, no current flows at small source-drain bias. In this regime the system is blocked, until V_{sd} exceeds the charging energy of the dot. There is a value of the gate voltage at which the probability of finding an electron on the dot is ½, at which point electrons hop on and off the dot at arbitrarily small values of the source-drain bias. Figure 7 shows a map of the differential conductance of the device, plotted as a gray scale, as a function of both V_{sd} and V_g. Multiple coulomb oscillations are observed, indicating that successive charges can be added or removed from the dot.

The characteristics of this device differ substantially from what would be

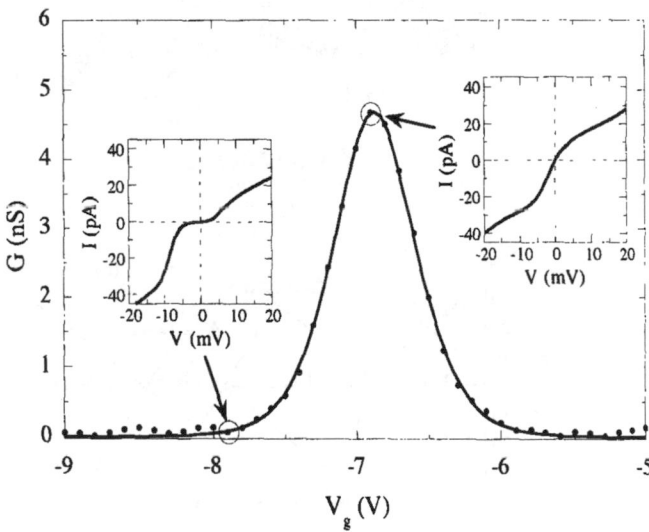

Figure 6. Current-voltage characteristics of the single nanocrystal single electron transistor device.

observed if the active element were metallic. For instance, the spacing of the successive charge states is not equal in energy. Note that on the right side of Figure 7, the Coulomb gap continues to grow with increasingly positive V_g. For even larger gate voltages than] shown in the figure (up to 40 V) the Coulomb gap exceeds 150 meV and there is no evidence for any more Coulomb oscillations. This behavior can be understood if electrons are being added to the valence band of the nanocrystal with increasingly positive V_g. In this case, for some sufficiently positive V_g, every state in the valence band is filled, and the next extended state of the nanocrystal lies in the conduction band, 2 eV higher in energy. In this interpretation, the large gap at positive V_g corresponds to a completely filled valence band. Starting on the right side of Figure 7c and decreasing V_g, the first Coulomb oscillation then corresponds to the removal of the first electron from, or, alternatively, the addition of the first hole to, the valence band. A further decrease in V_g adds additional holes. The addition energies for the 2nd, 3rd, and 4th holes are: $_2 = 14 \pm 2$ meV, $_3 = 29 \pm 3$ meV, and $_4 = 22 \pm 2$ meV.

This work represents a new type of spectroscopy for single nanocrystals.

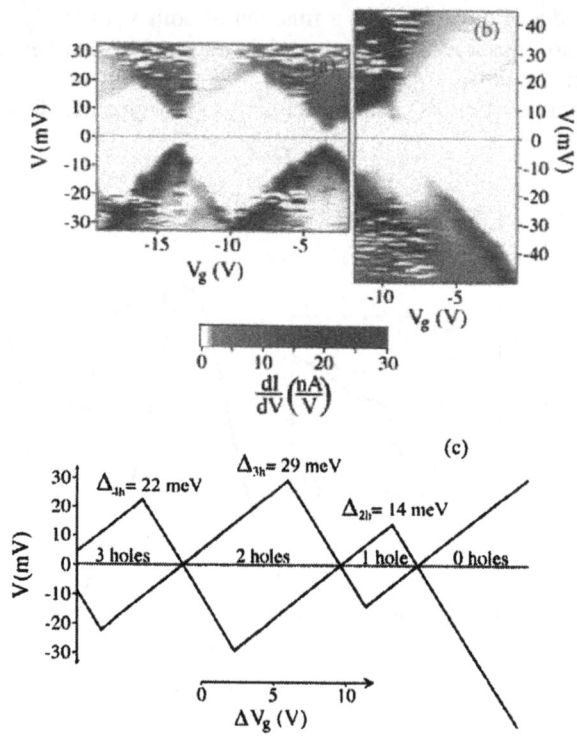

Figure 7. A plot of the differential conductance versus gate voltage and source drain bias shows the effect of adding multiple holes to the dot.

Unlike optical measurements, where electron-hole pairs are created, these measurements probe the energy for adding a single type of charge carrier. Future measurements will investigate how the ground state and excited state properties vary with the size, shape, and composition of nanocrystals as well as the composition of the leads. It is important to note that such devices are not yet of any practical use since their fabrication relies on the fortuitous placement of a nanocrystal fluctuations in local field modulate the gate voltage. Controlled techniques for the spatial organization of nanocrystals[41-44] are under development by a number of groups.

1.9 CONCLUDING REMARKS

Colloidal semiconductor nanocrystal samples have continued to advance rapidly. It is now possible to make large amounts of monodisperse II-VI and III-V nanocrystals, with a high degree of surface passivation, and with the ability to manipulate the environment. Such nanocrystals have now been incorporated into electrical devices, substantially extending the ability to understand and exploit them. In particular, composites of nanocrystals with polymers provide an excellent route for electrically accessing ensembles of nanocrystals; single nanocrystal electrical studies are also of current interest.

Acknowledgments: Many colleagues have contributed to the work described here, including Dr. Xiaogang Peng, Dr. Chia-Chun Chen, Amy Herhold, Mike Schlamp, Dr. Neil Greenham, David Klein, Andrew Lim, and Professor Paul McEuen. We thank the Department of Energy, the National Science Foundation, and the Office of Naval research for support.

References

[1] L. E. Brus, A. L. Efros, and T. Itoh, "Special Issue On Spectroscopy Of Isolated and Assembled Semiconductor Nanocrystals - Introduction," Journal Of Luminescence 70 (V70), R7-R8 (1996).

[2] H. Weller, "Optical Properties Of Quantized Semiconductor Particles," Philosophical Transactions Of the Royal Society Of London Series a-Mathematical Physical and Engineering Sciences 354 (1708), 757-766 (1996).

[3] T. Vossmeyer, G. Reck, B. Schulz et al., "Double-Layer Superlattice Structure Built Up Of Cd32s14(Sch2ch(Oh)Ch3)(36)Center-Dot-4h(2)O Clusters," Journal Of the American Chemical Society 117 (51), 12881-12882 (1995).

[4] Y. Wang, M. Harmer, and N. Herron, "Towards Monodisperse Semiconductor Clusters - Preparation and Characterization Of Similar-to 13-Angstrom Thiophenolate-Capped Cds Clusters," Israel Journal Of Chemistry 33 (1), 31-39 (1993).

[5] A. P. Alivisatos, "Semiconductor Clusters, Nanocrystals, and Quantum Dots," Science 271 (5251), 933-937 (1996).

[6] S. A. Harfenist, Z. L. Wang, M. M. Alvarez et al., "Highly Oriented Molecular Ag Nanocrystal Arrays," Journal Of Physical Chemistry 100 (33), 13904-13910 (1996).

[7] R. L. Whetten, J. T. Khoury, M. M. Alvarez et al., "Nanocrystal Gold Molecules," Advanced Materials 8 (5), 428+ (1996).

[8] C. P. Collier, R. J. Saykally, J. J. Shiang et al., "Reversible tuning of silver quantum dot monolayers through the metal-insulator transition," Science 277 (5334), 1978-1981 (1997).

[9] Shi Jing, S. Gider, K. Babcock et al., "Magnetic clusters in molecular beams, metals, and semiconductors," Science 271 (5251), 937-41 (1996).

[10] J. M. McHale, A. Auroux, A. J. Perrotta et al., "Surface energies and thermodynamic phase stability in nanocrystalline aluminas," Science 277 (5327), 788-791 (1997).

[11] H. Gleiter, "Nanostructured materials," Advanced Materials 4 (7), 474-81 (1992).

[12] H. Gleiter, "Nanostructured materials: state of the art and perspectives," Zeitschrift fur Metallkunde 86 (2), 78-83 (1995).

[13] Ph. Buffat and J.P. Borel, " Size effect on the melting temperature of gold particles," Physical Review A 13 (6), 2287-98 (1976).

[14] A. N. Goldstein, C. M. Echer, and A. P. Alivisatos, "Melting In Semiconductor Nanocrystals," Science 256 (5062), 1425-1427 (1992).

[15] C. B. Murray, D. J. Norris, and M. G. Bawendi, "Synthesis and Characterization Of Nearly Monodisperse Cde (E = S, Se, Te) Semiconductor Nanocrystallites," Journal Of the American Chemical Society 115 (19), 8706-8715 (1993).

[16]J. E. B. Katari, V. L. Colvin, and A. P. Alivisatos, "X-Ray Photoelectron Spectroscopy Of Cdse Nanocrystals With Applications to Studies Of the Nanocrystal Surface," Journal Of Physical Chemistry 98 (15), 4109-4117 (1994).

[17]M. A. Olshavsky, A. N. Goldstein, and A. P. Alivisatos, "Organometallic Synthesis Of Gaas Crystallites Exhibiting Quantum Confinement," Journal Of the American Chemical Society 112 (25), 9438-9439 (1990).

[18]O. I. Micic and A. J. Nozik, "Synthesis and Characterization Of Binary and Ternary Iii-V Quantum Dots," Journal Of Luminescence 70 (V70), 95-107 (1996).

[19]A. A. Guzelian, J. E. B. Katari, A. V. Kadavanich et al., "Synthesis Of Size-Selected, Surface-Passivated Inp Nanocrystals," Journal Of Physical Chemistry 100 (17), 7212-7219 (1996).

[20]A. A. Guzelian, U. Banin, A. V. Kadavanich et al., "Colloidal Chemical Synthesis and Characterization Of Inas Nanocrystal Quantum Dots," Applied Physics Letters 69 (10), 1432-1434 (1996).

[21]S. H. Tolbert and A. P. Alivisatos, "High-Pressure Structural Transformations In Semiconductor Nanocrystals," Annual Review Of Physical Chemistry 46 (V46), 595-625 (1995).

[22]Chia-Chun Chen, A. B. Herhold, C. S. Johnson et al., "Size dependence of structural metastability in semiconductor nanocrystals," Science 276, 398-401 (1997).

[23]M. G. Bawendi, M. L. Steigerwald, and L. E. Brus, "The Quantum Mechanics Of Larger Semiconductor Clusters (Quantum Dots)," Annual Review Of Physical Chemistry 41 (V41), 477-496 (1990).

[24]Ulrike Woggon, Optical Properties of Semiconductor Quantum Dots (Springer, Berlin, 1996).

[25]M. A. Hines and P. Guyotsionnest, "Synthesis and Characterization Of Strongly Luminescing Zns-Capped Cdse Nanocrystals," Journal Of Physical Chemistry 100 (2), 468-471 (1996).

[26]X. G. Peng, M. C. Schlamp, A. V. Kadavanich et al., "Epitaxial growth of highly luminescent CdSe/CdS core/shell nanocrystals with photostability and electronic accessibility," Journal Of the American Chemical Society 119 (30), 7019-7029 (1997).

[27]Y. N. C. Chan, R. R. Schrock, and R. E. Cohen, "Synthesis Of Single Silver Nanoclusters Within Spherical Microdomains In Block Copolymer Films," Journal Of the American Chemical Society 114 (18), 7295-7296 (1992).

[28]C. C. Cummins, R. R. Schrock, and R. E. Cohen, "Synthesis Of Zns and Cds Within Romp Block Copolymer Microdomains," Chemistry Of Materials 4 (1), 27-30 (1992).

[29]M. Antonietti and C. Goltner, "Superstructures of functional colloids: Chemistry on the nanometer scale," Angewandte Chemie-International Edition In English 36 (9), 910-928 (1997).

[30]B. O'Regan and M. Gratzel, "A Low-Cost, High-Efficiency Solar Cell Based On Dye-Sensitized Colloidal Tio2 Films," Nature 353 (6346), 737-740 (1991).

[31]N. C. Greenham, X. G. Peng, and A. P. Alivisatos, "Charge separation and transport in conjugated-polymer/semiconductor-nanocrystal composites studied by photoluminescence quenching and photoconductivity," Physical Review B-Condensed Matter 54 (24), 17628-17637 (1996).

[32]M. Schlamp, Xiaogang Peng, and A. P. Alivisatos, "Improved efficiencies in light emitting diodes made with CdSe (CdS) core/shell type nanocrystals and a semiconductor polymer," Journal of Applied Physics in press (1997).

[33]V. L. Colvin, M. C. Schlamp, and A. P. Alivisatos, "Light-Emitting Diodes Made From Cadmium Selenide Nanocrystals and a Semiconducting Polymer," Nature 370 (6488), 354-357 (1994).

[34]B. O. Dabbousi, M. G. Bawendi, O. Onitsuka et al., "Electroluminescence From Cdse Quantum-Dot Polymer Composites," Applied Physics Letters 66 (11), 1316-1318 (1995).

[35]D. L. Klein, P. L. McEuen, J. E. B. Katari et al., "An Approach to Electrical Studies Of Single Nanocrystals," Applied Physics Letters 68 (18), 2574-2576 (1996).

[36]S. A. Blanton, A. Dehestani, P. C. Lin et al., "Photoluminescence Of Single Semiconductor Nanocrystallites By Two-Photon Excitation Microscopy," Chemical Physics Letters 229 (3), 317-322 (1994).

[37]S. A. Blanton, M. A. Hines, M. E. Schmidt et al., "Two-Photon Spectroscopy and Microscopy Of Ii-Vi Semiconductor Nanocrystals," Journal Of Luminescence 70 (V70), 253-268 (1996).

[38]M. Nirmal, B. O. Dabbousi, M. G. Bawendi et al., "Fluorescence Intermittency In Single Cadmium Selenide Nanocrystals," Nature 383 (6603), 802-804 (1996).

[39]S. A. Empedocles, D. J. Norris, and M. G. Bawendi, "Photoluminescence Spectroscopy Of Single Cdse Nanocrystallite Quantum Dots," Physical Review Letters 77 (18), 3873-3876 (1996).

[40]D. L. Klein, R. Roth, A. K. Lim et al., "A single-electron transistor made from a cadmium selenide nanocrystal," Nature in press, in press (1997).

[41]A. P. Alivisatos, K. P. Johnsson, X. G. Peng et al., "Organization Of Nanocrystal Molecules Using Dna," Nature 382 (6592), 609-611 (1996).

[42]C. A. Mirkin, R. L. Letsinger, R. C. Mucic et al., "A Dna-Based Method For Rationally Assembling Nanoparticles Into Macroscopic Materials," Nature 382 (6592), 607-609 (1996).

[43]P. C. Ohara, J. R. Heath, and W. M. Gelbart, "Self-assembly of submicrometer rings of particles from solutions of nanoparticles," Angewandte Chemie-International Edition In English 36 (10), 1078+ (1997).

[44]T. Vossmeyer, E. DeIonno, and J. R. Heath, "Light-directed assembly of nanoparticles," Angewandte Chemie-International Edition In English 36 (10), 1080-1083 (1997).

2 Injection and Transport of Spin Coherence in Semiconductors

D. D. Awschalom[1] and N. Samarth[2]
[1]*Department of Physics, University of California, Santa Barbara, CA 93106, USA and* [2]*Department of Physics, Pennsylvania State University, University Park, PA 16802, USA*

ABSTRACT

While conventional electronic devices rely on charge for the storage and transport of information, the use of spin for such purposes may lead to new paradigms in quantum electronics with improved speed and qualitatively different functionality. A generic requirement of these new systems is the storage and manipulation of classical and/or quantum spin information. Femtosecond-resolved optical techniques are used to create a superposition of the electronic basis spin states defined by an applied field, and to follow the phase, amplitude, and location of the precessing spin population. Here we discuss studies in both magnetically- and electronically-doped quantum structures and thin films, demonstrating a rich variety of coherent spin phenomena in these systems.

1.1 INTRODUCTION

While contemporary solid state devices rely on the transport and storage of electronic charge, it is interesting to consider the possibility of using the electronic *spin* as a basis for future technologies. Recently, the fields of quantum electronics and micromagnetics have been converging towards a new discipline known as "magnetoelectronics," which focuses on low dimensional electronic systems that display magnetically-driven spin-dependent phenomena. On the one hand, quantum electronics has been remarkably successful at exploiting nanofabrication techniques to establish quantized energy levels, where charge manipulation produces a variety of electronic and optical devices. In parallel, research in micromagnetism has cleverly used miniaturization to produce submicron ferromagnetic structures whose switching and stability are controlled to create magnetic storage media. The merging of these two areas enables magnetoelectronic phenomena where the spin of quantum confined charge carriers is controlled using local magnetic fields. While there are several recent examples of systems where magnetic nanostructures are used in conjunction with semiconductor heterostructures to direct electron flow (Prinz, 1995; Gider et al., 1996, Tornow et al., 1996; Prinz, 1990), the integration of magnetic and electronic quantum structures also offers opportunities to probe qualitatively new physics by obtaining additional *dynamical* information about the full quantum mechanical nature of electronic states in reduced geometries (Awschalom and Samarth, 1993). The eventual goal of such studies is to establish, store, and manipulate the *coherence* of electronic and magnetic spins in solid state systems.

Here we describe a program that involves the concurrent development of "spin-engineered" II-VI magnetic semiconductor quantum wells, III-V semiconductors, and time-resolved optical spectroscopies capable of probing electronic and magnetic spin behavior within the relevant temporal and spatial scales (Awschalom and Samarth, 1998). The electronic dynamics are probed by optically exciting a coherent superposition of basis spin states defined by an applied field, and then following the evolution of this "spin packet" as angular momentum is transferred from a circularly polarized photon pulse to charge carriers and – in the case of magnetic semiconductors - eventually to a sublattice of magnetic ions (Crooker et al., 1996; Crooker et al., 1997; Crooker et al., 1995). In this latter case, terahertz spin precession of photoinjected electrons is observed, accompanied by a rapid spin-relaxation of holes. Subsequently, the experiments demonstrate the coherent transfer of angular momentum to the magnetic sublattice where the perturbed ions undergo free-induction decay at microwave frequencies, enabling time-domain all-optical electron spin resonance measurements in magnetic nanostructures. Measurements in electronically-doped nonmagnetic structures reveal the surprising discovery of extended electronic spin memory, where nanosecond spin precession times are observed at room temperature (Kikkawa et al., 1997). Finally, studies in electrically-doped bulk III-V GaAs materials show a dramatic extension of these time scales, suggesting that this physics may be present in a wide variety of solid state systems (Kikkawa and Awschalom, 1998). This has led to the recent demonstration of macroscopic spin transport (Kikkawa and Awschalom, 1999).

1.2 MAGNETIC SEMICONDUCTOR HETEROSTRUCTURES

It is important to establish a flexible material system for these experiments which optimally combines elements of semiconductor physics and magnetism so as to enable direct study of the dynamical evolution of angular momentum in a low dimensional electronic system, where electrons can be manipulated into specific locations using well-defined bandgap engineering. To this end, we have developed "digital magnetic heterostructures" (DMH) in which interactions between localized magnetic spins and their overlap with quantum-confined electronic states is tailored through a controlled digital distribution of two-dimensional (2D) magnetic layers (Crooker et al., 1995). This class of quantum structures, shown in figure 1, provides a widely tunable two-level electronic spin system with qualitatively different dynamical interactions than those seen in traditional diluted magnetic semiconductor (DMS) alloys. Structures are grown on (100) GaAs substrates using molecular beam epitaxy (MBE) of II-VI semiconductors, and consist of 120Å $Zn_{.77}Cd_{.23}Se/ZnSe$ single quantum wells (QWs) containing different local concentrations of Mn^{2+} ions. In the digital structures, a fixed total amount of MnSe (3 monolayers) is incorporated into the well in a series of equispaced planes, allowing direct control over the average distance and coupling between neighboring Mn^{2+} spins. The QWs contain a single three-monolayer barrier of MnSe (1x3ml), three equispaced single monolayers of MnSe (3x1ml), 24 one-eighth monolayers (24x1/8ml), or a nonmagnetic control (NM). A fifth sample contains a single four-monolayer barrier of $Zn_{.90}Mn_{.10}Se$ (4ml 10%). The growth rates of the various components of the heterostructures are determined to within 5% using *in situ* electron diffraction studies. X-ray diffraction studies of corresponding multilayer samples indicate that the interdiffusion profile has a relatively narrow width (~1 ml) during growth, giving local quasi-2D Mn^{2+} concentrations of 100%, 50%, 8% and 10% respectively.

Low-temperature static photoluminescence (PL) and absorption studies are performed using the Faraday configuration in B < 8 T. All samples reveal sharp heavy-hole (*hh*) exciton peaks with small Stokes shifts (~ 3 meV) whose FWHM (~6 meV) are attributed to inhomogeneous broadening caused by nonmagnetic alloy and well width fluctuations. An applied magnetic field splits the exciton into lower (spin-down, $S_z=+1$) and higher energy (spin-up, $S_z=-1$) states due to the sp-d exchange interaction resulting from overlap of the electron and heavy hole wavefunctions and the distributed Mn spins. The polarization-resolved magnetoabsorption from spin-up and spin-down excitons is shown in figure 1. The PL linewidths and corresponding absorption oscillator strengths are independent of applied fields and are nearly the same as from the control structure, indicating little change in the electronic confinement. The spin-split absorption spectra illustrate how DMH samples can provide prototypes of field-tunable two level spin systems, in which the spin-splittings are significantly larger than the inhomogeneous absorption linewidths even in modest magnetic fields (1T). Successive division of the magnetic layers produces two clear results: increased penetration of the wavefunction into the Mn barriers which increases the g-factor (figure 1), and a reduction of the wavefunction curvature causing the luminescence peaks to shift to lower energy. Both effects reflect significant changes in the Mn distribution. More subtle effects relate directly to changes in the local magnetic environment, showing a superlinear dependence of the g-factor as a function of the electronic overlap with the magnetic regions, and is likely due to an increase in the number of uncompensated paramagnetic spins.

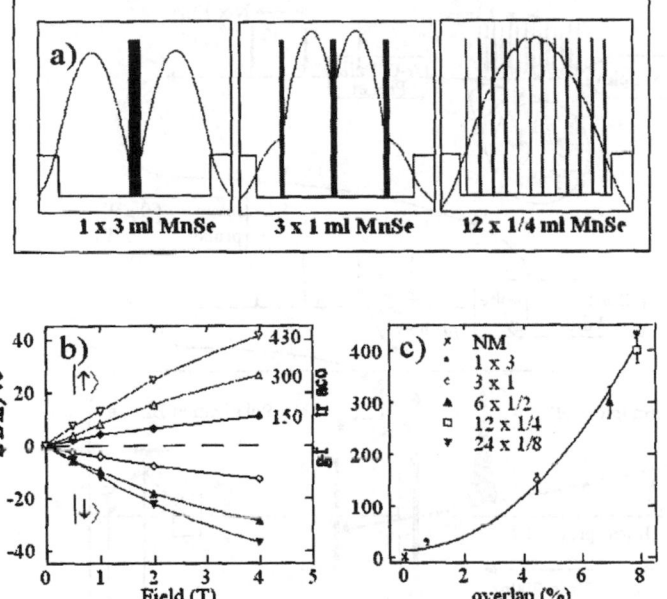

Figure 1. (a) Schematic diagram of the conduction band energy profiles and electron wavefunctions. (b) Zeeman-splittings and g-factors of the exciton absorption peaks as a function of magnetic field at 5K. Lines are guides to the eye. (c) Low field g-factors vs. the calculated wavefunction overlap with the MnSe barriers.

1.2.1 Magneto-optical spectroscopy

The samples are mounted in a magneto-optical cryostat in transverse fields $(H \parallel \hat{x})$, aligning the Mn^{2+} moments along the plane of the quantum well and probed as shown in figure 2. Spin-polarized electrons and holes, initially oriented normal to the plane of the well $(S^{e,h} \parallel \hat{z})$, are optically injected by 120fs pump pulses of circularly polarized light (~50 µW) tuned to the peak of the *hh* resonance. The 76 MHz repetition rate of the laser is reduced with an acousto-optic pulse picker to allow for complete recovery of the magnetic system between exciting pulses. The net magnetization of the sample parallel to $\hat{z}$ is measured via the Faraday rotation imparted to a weak, time-delayed, linearly-polarized probe pulse (~1 µW). In analogy with the familiar Wheatsone bridge used in electronic circuits, an optical polarization bridge measures the pump-induced changes in the Faraday rotation of the probe with sub-millidegree sensitivity within a spot size of typically ~50 µm (Crooker et al., 1995). Note that this technique measures M_z, t component of magnetization along the direction of the probe beam.

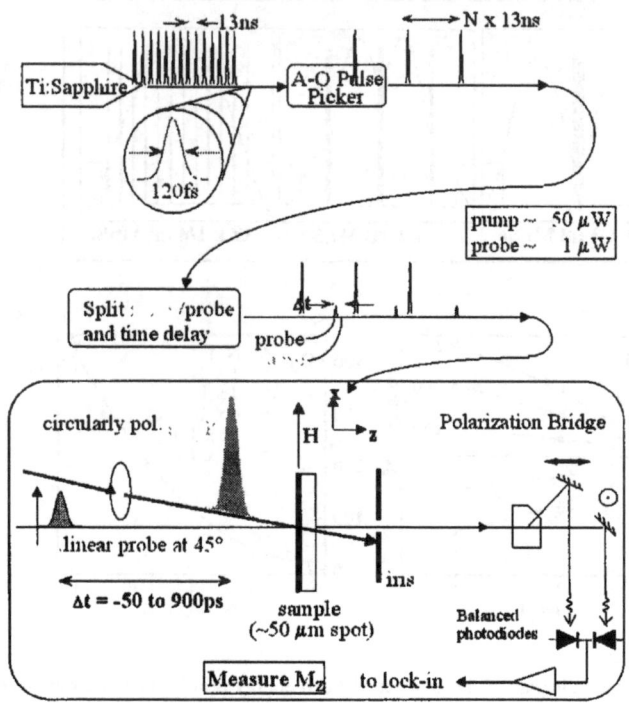

Figure 2. Schematic diagram of the time-resolved Faraday rotation experiment. The pulse-picker reduces the repetition rate of the laser. The optical bridge serves as a null detector for changes in the rotation of polarization.

1.2.2 Electron and hole dynamics

Figure 3(a) shows the Faraday rotation following injection of spin-polarized electrons ($S_z^e = \pm 1/2$) and holes ($S_z^h = \mp 3/2$) into the 4ml 10% magnetic quantum well. The Faraday rotation on these short timescales reflects the component of net carrier spin oriented along the $\hat{z}$-axis and displays the expected reversal of sign upon pumping with the opposite sense of circular polarization. In zero applied field, the observed superposition of two exponential decays arises from the spin relaxation of the electrons and holes, which then recombine on a longer timescale. As discussed below, the hole spins relax quickly (<5ps) in these zinc-blende structures despite the removal of the light hole-heavy hole degeneracy by strain, while electron spin scattering is significantly slower. Similar behavior has been predicted and observed in unstrained GaAs quantum wells, with the rapid hole spin relaxation attributed to valence band mixing (Bar-Ad and Bar-Joseph, 1992; Uenoyama and Sham, 1990).

This monotonic behavior is significantly modified upon application of a transverse magnetic field, H_x. Pronounced oscillations appear in the measured Faraday rotation resulting from the precession of the net electron spin $\langle S_z^e \rangle$ about the orthogonal field, $B_x = \mu_o(H_x + M_x^{Mn})$. This field-tunable Larmor precession frequency can be several THz in magnetic semiconductor systems due to the strong s-d exchange between the conduction band and the local Mn^{2+} spins. The electrons are quantized along the field axis ($S_x^e = \pm 1/2$); thus photoinjecting electron spins with definite S_z^e corresponds to pumping a coherent superposition of the spin-split electron states $\pm S_x^e$. This superposition "beats" at the frequency determined by the electron Zeeman splitting, $\Delta E = g_{eff}^e \mu_B H_x$. Consequently, the Larmor frequency, $\Omega_L = \Delta E / h$, directly provides the effective g-factor g_{eff}^e and spin-splitting ΔE of the electrons alone, as recently found in luminescence experiments on non-magnetic GaAs (Herberle et al., 1994). Figure 3(b) shows the field dependence of Ω_L and ΔE in the 24x1/8ml sample at different temperatures. As expected, Ω_L tracks the magnetization of the sample, which follows a Brillouin function arising from the alignment of the paramagnetic Mn^{2+} spins (Furdyna and Kossut, 1988). Also shown is the corresponding *exciton* Zeeman splitting, reduced by a factor of 5.7, as measured by the spin-splitting of the *hh* exciton absorption peaks in longitudinal magnetic fields (Crooker et al., 1995). These data clearly support identification of the precession signal as arising from the electrons alone - reflectivity measurements in bulk $Zn_{1-x}Mn_xSe$ indicate nearly the same ratio (5.8) between exciton and electron g-factors (Twardowski et al., 1983).

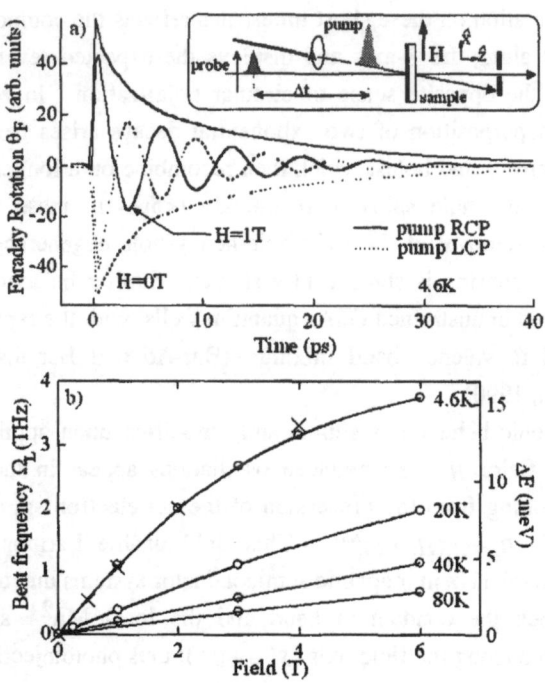

Figure 3. (a) The pump-induced Faraday rotation, measuring net carrier spin $<S_z>$, in transverse fields H=0, 1T for pumping with right- (solid) and left- (dashed) circularly polarized light in the quantum well containing the 4-monolayer $Zn_{.90}Mn_{.10}Se$ barrier. Inset : Schematic of experiment. Pump/probe polarizations are as drawn. (b) Electron precession frequency and corresponding electron Zeeman splitting at 4.6K in the 24x1/8 ml sample vs. applied field. Crosses (X) are exciton splitting scaled down by factor 5.7.

At early times the measured Faraday rotation is fit well by the sum of an exponentially-decaying cosinusoidal oscillation (electrons) and a faster purely exponential decay (figure 4(a)), which we identify with the rapid spin relaxation of the photoinjected hole population. The holes do not precess; their angular momenta are constrained to lie along the growth axis ($c\|\hat{z}$) due to strain and quantum confinement, resulting in a vanishing *hh* spin-splitting in transverse magnetic fields (for B<8T) (Martin et al., 1990; Kuhn-Heinrich and Ossau, 1995). This may be viewed using an atomic orbital picture shown in figure 4(b) as follows: recall that in this system electrons are s-states and that the holes are p-states. The *hh-lh* transitions are split by quantum confinement, where the *lh* energy is unfavorable with respect to the *hh* energy due to excess penetration into the barrier. The *hh* state (l=1) is a combination of p_x and p_y orbitals that form the familiar "doughnut" spatial mode, resulting in the angular momentum pointing along the z-direction. Thus, in contrast to the electron, the *hh* cannot precess without an admixture of the *lh* state. Finally, the spin-orbit interaction aligns the hole spin with the angular momentum perpendicular to the quantum well plane, yielding rapid monotonic spin relaxation. This ability to clearly differentiate between the evolution of the net electron and hole spin enables detailed study of the role of applied field and magnetic environment upon the individual populations' spin relaxation.

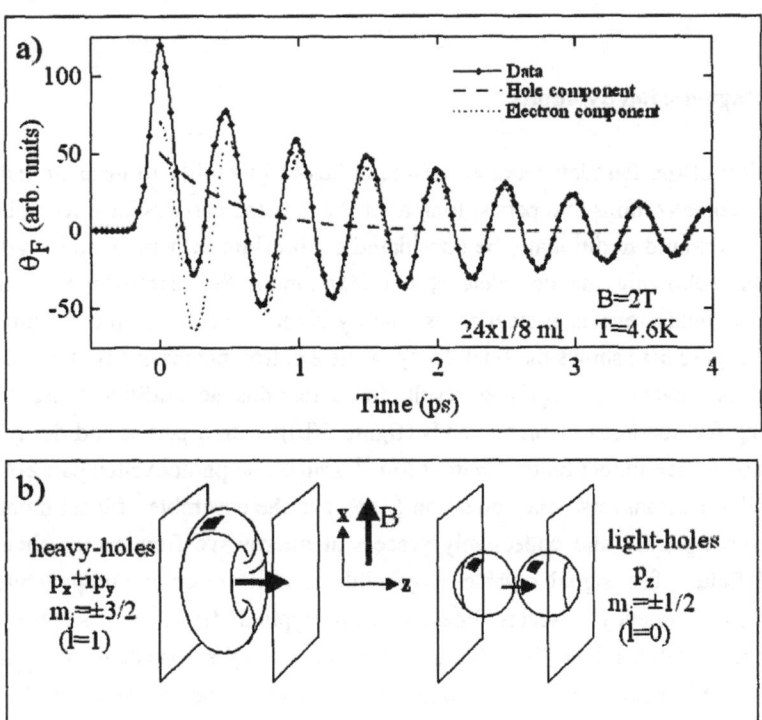

Figure 4. (a) Induced Faraday rotation (solid) in the 24x1/8 ml sample, showing the rapid initial decay of the holes superimposed on the electron precession. (b) Cartoon showing the atomic orbital angular momenta for the heavy-hole and light-hole states within a quantum-confined geometry. Note that the

higher-energy light-hole state penetrates the surrounding barriers compared to the preferred lower-energy heavy-hole state.

The electron oscillations are directly analogous to the free-induction decays of transverse nuclear spin which follow a $90°$ tipping pulse in NMR . In applied fields, we measure a transverse spin relaxation T_2^* of the photoinjected electrons which is related to the longitudinal spin relaxation time (T_1) and the homogeneous dephasing time (T_2) in magnetic resonance by the relation $1/T_2^* \cong 1/(2T_1) + 1/T_2'$. While the precise contribution from the two effects is still under investigation, it is clear that the increase in spin relaxation rate with field is much larger for those samples with little or no coupling to the Mn moments. In contrast to data drawn from the Hanle effect, these measurements provide a direct and simple temporal characterization of spin scattering, and offer a new method of time-domain electron spin resonance in undoped semiconductor quantum structures.

1.2.3 Magnetic ion dynamics

The utility of the Faraday rotation technique lies in its ability to measure induced sample magnetizations that persist long after the injected carriers have recombined, a state that would traditionally be considered equilibrium in nonmagnetic systems. However, following the complete spin relaxation of the electrons, the data in transverse fields reveals a smaller oscillatory signal with a completely different period. Figure 5(a) shows the final decay of the electron precession in the 4ml 10% sample and reveals a surprising result: there remains an additional oscillation, persisting for hundreds of picoseconds (figure 5(b)), with a period and decay time implying the free induction of coherent Mn^{2+} spins. The photoexcited carriers have imparted a net transverse magnetization $\langle \Delta M_z^{Mn} \rangle$ to the ensemble of local moments, which subsequently and collectively precess at microwave frequencies about the applied field. The signal reverses sign with opposite circular pump, while the precession frequency   scales linearly with applied field,   is sample- and temperature- independent (2-100K), and corresponds to that expected for $g_{Mn} = 2.0$. The observed decay time of these magnetic oscillations agrees well (±10%) with ESR measurements of the dephasing time (T_2^{Mn}) of Mn^{2+} spins in bulk $Zn_{.90}Mn_{.10}Se$, (Samarth and Furdyna, 1988) confirming the realization of an all-optical time-domain spin resonance experiment that is easily capable of probing small numbers of spins.

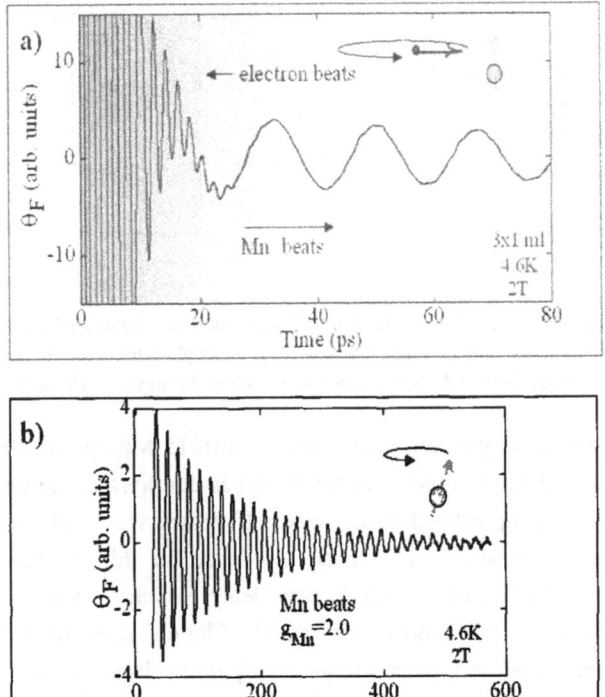

Figure 5. (a) Induced Faraday rotation, showing the final oscillations of the electrons superimposed on an induced precession of the Mn spins in the sample containing the 4ml $Zn_{.90}Mn_{.10}Se$ barrier. The highlighted area represents the PL lifetime of optically-excited carriers in this sample. (b) Evolution of the Mn precession, showing longlived free-induction decay.

A possible mechanism for initiating the observed spin resonance signal lies in the impulsive *coherent* rotation of Mn^{2+} moments about the transient exchange field ($H_{exch} \| \hat{z}$) generated by the hole spins. To first order, the electrons can be excluded from consideration since any contribution averages away due to their much faster precession and weaker coupling to the Mn sublattice. As shown in figure 6, the sample magnetization $\vec{M}^{Mn}$, oriented initially along the applied field H_x, is rotated away from the $\hat{x}$-axis upon application of a torque $\| \hat{y}$ by the strong exchange field of the holes, which persists for the hole spin relaxation time, acting on each ion. A similar process has been invoked to explain Raman data in

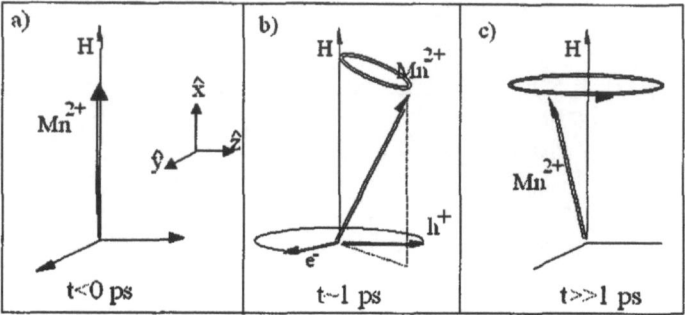

Figure 6. Model for coherent rotation of the Mn sublattice about the transient hole exchange field. The Mn spins are (a) oriented initially along the $\hat{x}$ axis, of the applied field, are (b) tipped into $\hat{y}$ by the transient hole exchange field, and (c) remain to precess about the applied field for long times.

magnetic semiconductor quantum wells, where up to 15 Mn spin-flip Stokes lines are observed (Stühler et al., 1995). After the holes equilibrate, the perturbed Mn moment, which has been rotated by up to a half degree from the x-axis in these experiments, then precesses freely. Strong evidence for this mechanism is seen from extrapolating the Mn beats back to zero delay, which shows the oscillations build up sinusoidally as predicted in this model. The effect of the hole exchange field is directly analogous to the radiofrequency tipping pulses used in NMR studies to initiate free induction decays in nuclear moments (Fukushima and Roeder, 1981). In essence, a single hole is tipping a large number of Mn spins within the spatial extent of its wavefunction.

This model of coherent rotation about the exchange field of the holes predicts several trends which are readily confirmed by experiment. The first prediction involves the measured amplitude of the Mn beats which is expected to be proportional to the net sample magnetization and thus increase from zero with the applied field. In zero field there is no net magnetization, and no Mn beats are in fact observed. Moreover, this result rules out the possibility of polarization of the Mn sublattice by direct spin-flip scattering with the carriers, which would be expected to induce a longlived magnetization along $\hat{z}$ even in zero field. A second prediction of the model is that the hole lifetime will affect the tipping angle of the magnetic ions. This has been tested by simply repeating these measurements in quantum structures with varying well widths (Crooker et al., 1997). The normalized magnitude of the Mn oscillation signal is largest in the narrowest quantum well, and decreases with increasing well-width. This trend is consistent with a scenario of increased hole stability provided by the stronger quantum confinement in the narrower wells. The observed behavior is also qualitatively consistent with recent calculations that show a decrease in the number of Raman spin flip resonances with increasing well width (Kavokin and Merkulov, 1997). A longer-lived hole exchange field in narrower wells tips the Mn^{2+} moments further away from the axis of the applied field, leading to larger amplitude beats. Lastly, because a rotation of the Mn^{2+} spins about the hole exchange field initially tips the moments into the y-axis, the Mn^{2+} precession signal should commence as a sinusoid. That is, if the Mn^{2+} beats are fit to an exponentially decaying sinusoid, $Ae^{-t/\tau}\sin(\Omega_{Mn}t + \phi)$, the oscillatory signal should extrapolate back to zero amplitude (or alternatively, zero phase) at zero time delay. Figure 7 shows the initial phase of the Mn beats at 5K as a function of time. In field, an extrapolation of the Mn beats back to zero time delay indicates that the Mn oscillation signal is initially positive, with negative slope. However, this effect is artificial. Careful analysis of the signal as compared to the fit at times near zero reveals an interesting phenomenon: the fit and the data develop an increasing phase shift as one looks 'backward' in time towards zero delay, as shown in figure 7. The Mn^{2+} spins initially begin precessing slowly, then accelerate to a higher frequency commensurate with g=2.01. The Mn^{2+} moments experience an initially reduced magnetic field which grows asymptotically to the applied magnetic field. We believe that during the ~30ps over which this effect is seen to occur, the Mn^{2+} spins are experiencing the applied field reduced by the demagnetization field of the spin-relaxed excitons. Recall that the exciton recombination times are much longer than the spin lifetimes, and thus they are still present in the sample to ~50 ps. The excitons have relaxed to their lowest energy state and are preferentially oriented antiparallel to the applied field and are ostensibly invisible to the Faraday rotation probe. However, these oriented excitons generate a demagnetizing field in the sample, and as they recombine the strength of this demagnetizing field decays to zero, whereafter the Mn spins precess faster about the 'bare' applied field. Thus the presence of the 'invisible' excitons is indirectly inferred through the precession frequency of the Mn moments, and we conclude that the measured Mn spins do indeed commence as a sinusoid, in support of the model for coherent rotation.

Note that one may use this optical generation and detection of free-induction decays in the embedded Mn^{2+} ions to explore the spin resonance of diluted magnetic planes at the monolayer level. The temperature-dependent dephasing times (T_2^{Mn}) of the Mn moments may be measured for varying spatial distributions, as well as the variation of T_2^{Mn} with temperature and spin distribution. We find this to be in qualitative agreement with exchange narrowing models

(Samarth and Furdyna, 1988), which relate the transverse relaxation time to anisotropic and isotropic spin-spin interactions, as well as to static and dynamic spin-spin correlations. Note that while the 24x1/8ml sample and the 3x1ml sample contain an identical number of Mn spins, the different local spin densities (~8% and ~50%) are clearly evidenced in the disparate decay times which are measured with this technique (230ps and 70ps, respectively, at 4.6K and 3T), highlighting the crucial role of nearest neighbor spin-spin interactions in determining the transverse relaxation process. An important difference between the all-optical ESR measurement carried out here and conventional frequency domain ESR is the capability of continuously measuring the variation of T_2^{Mn} with magnetic field. For instance, at low temperatures, the Mn spin dephasing rate, $(T_2^{Mn})^{-1}$, increases dramatically with field for the sample with the highest local Mn density.

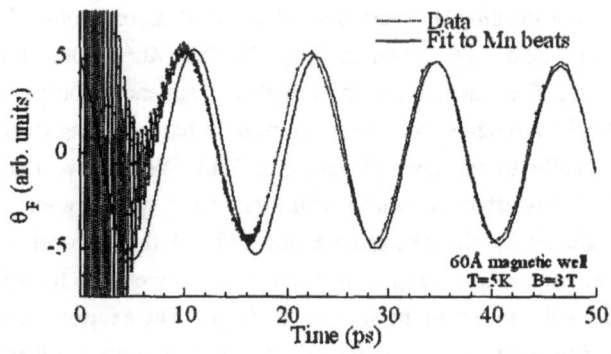

Figure 7. Dynamic phase-shifting of the Mn^{2+} free-induction signal with respect to the fixed frequency of the fit. The phase shift at short times is due to the presence of "invisible" carriers which generate a demagnetizing field. The fit to the data is excellent for all times >35 ps.

1.3 NONMAGNETIC ELECTRONICALLY DOPED STRUCTURES

Clearly it is of great scientific and technological interest to develop an understanding of these dephasing processes in order to extend spin coherence times in semiconductors. For example, recent discoveries that spin ensembles may be used collectively as single quantum elements have renewed optimism that coherent electronics may eventually be realized as a basis for computation (*Gershenfeld and*

Chuang, 1997). All the systems discussed thus far are nominally electrically insulating where the decay of electronic spin precession occurs within time scales of hundreds of picoseconds – except in magnetic semiconductors, where scattering off magnetic impurities induces decoherence in roughly ten picoseconds. In an effort to explore this phenomena in more conventional semiconductor nanostructures, we apply these methods used in the insulating magnetic quantum structures to electrically-doped nonmagnetic heterostructures.

1.3.1 Two-dimensional electron gases

Here we employ these time-resolved magneto-optical techniques to initiate and monitor electronic spin precession in modulation-doped II-VI semiconductor quantum wells which form two-dimensional electron gases (2DEG) at low temperatures (Kikkawa et al., 1997). Remarkably, we find that the presence of a 2DEG sustains this dynamical spin polarization nearly three orders of magnitude longer than in insulating samples at low temperatures. Moreover, the spin lifetime surpasses the recombination lifetime by 1-2 orders of magnitude, suggesting that the 2DEG acquires a net polarization either through energy relaxation of spin-polarized electrons or through angular momentum transfer within the electronic system. Our studies further show that these nanosecond spin lifetimes persist to room-temperature. Measurements on bulk II-VI epilayers reveal similar effects, even for low doping levels, and demonstrate that these phenomena are not restricted to quantum-confined electronic systems. These unexpected findings demonstrate that external contributions to electronic spin decoherence are subtantially reduced in these solid state systems.

Each 2DEG sample contains a strained 10.5 nm $Zn_{1-x}Cd_xSe$ QW (x~0.20) with Cl-doped ZnSe barriers, and is grown by molecular beam epitaxy on a semi-insulating (100) GaAs substrate. Samples include symmetrically-doped or asymmetrically-doped QWs with ZnSe spacer layers separating the QW from the n-type ZnSe doping layers, and a ZnSe buffer layer which separates the QW structure from the GaAs substrate. The details of these samples are given elsewhere. All 2DEG samples show a clear integer quantum Hall effect and Shubnikov-de Haas oscillations in transport measurements carried out at T < 4.2 K and in fields B > 2 T; low field Hall measurements are used to determine the 2DEG sheet densities and mobilities. Finally, we perform all of these studies in a reflection geometry in order to avoid undue strain or damage to these heterostructures from etching or other materials processing.

Figure 8(a) shows the low-temperature Kerr response of a 2DEG to a circularly polarized pump pulse in the presence of a 4 T in-plane magnetic field. Hundreds of oscillations in the Kerr rotation (individually resolved in the expanded inset) represent the precession of the injected electron spins about the applied field. This precession arises from the quantum beating of energy-split spin states in the conduction band at a frequency ν which is directly proportional to their spin splitting, $\Delta E = h\nu$, along the direction of the applied field. The deduced g-factors, $g = \Delta E/\mu_B H$, do not vary within the range of excitation powers and magnetic fields studied here and are only weakly temperature dependent below ~100 K. Thus a single value may be used to characterize the low temperature splittings in each sample, which we find to be nearly identical (g~1.1) for both insulating and doped QW structures. Clearly the addition of n-type dopants has extended the spin lifetimes by three orders of magnitude. We conclude that optical spin injection leads to a persistent spin polarization the 2DEG long after the carriers have recombined.

Most striking is the contrast between the response of the 2DEG and that of the insulating samples. The envelope of the 2DEG spin polarization evolves on two time scales: a fast decay over the first 100 ps followed by a slower, secondary decay over several nanoseconds. An insulating QW control sample as well as other undoped samples exhibit a much faster decay of the Kerr rotation, and only one resolved oscillation occurring at fields of 4T (Figure 8 (b)). A strength of this all-optical technique lies in its ability to directly resolve these decay profiles over both picosecond and nanosecond time intervals. Methods such as frequency-domain electron spin resonance (ESR) typically infer such dynamical behavior from absorption linewidths. To understand the significance of the Kerr response, we note that because θ_K is sensitive to the net magnetization of the electrons, the oscillation decay probes not only the number of precessing spins, but also their directional alignment. The former contributes to the homogeneous spin lifetime, T_2, in a transverse field, while the latter can be influenced by inhomogeneous effects such as local field variations or an energy dependent g-factor which introduce a progressive "dephasing" within the spin population. We denote the observed decay time as T_2^*, and note that this may also include contributions from spin diffusion away from the probe area ~ $(30 \ \mu m)^2$. As has been discussed elsewhere (Kikkawa et al., 1997), we do not believe that dephasing contributes to T_2^*.

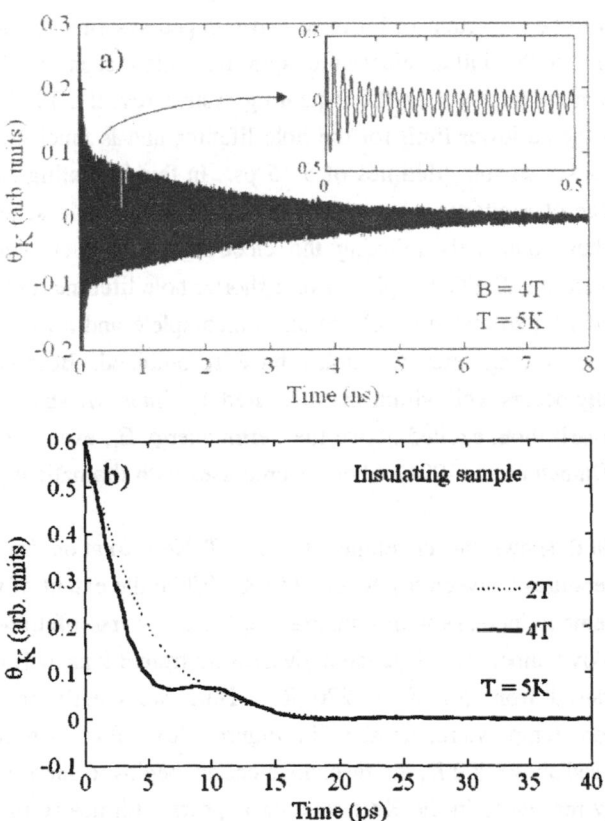

Figure 8. (a) Time-resolved Kerr rotation of a 2DEG with doping $n=2\times10^{11}$ cm^{-2} and mobility $\mu=6800$ cm^{2}V^{-1}s^{-1} at T=5K, B=4T, and an $E_{ex}=$ 2.70 eV excitation producing 5×10^{10} electrons cm^{-2}/pulse. The inset details the oscillations over the initial decay period. Axis labels remain unchanged. (b) the same as (a) for an insulating sample at T=5K and B=2 and 4T, with 3×10^{11} electrons/cm^{-2} per pulse and $E_{ex}=2.63$ eV. Note the difference in time scales between (a) and (b).

To better understand these decay profiles, we monitor carrier populations through time-resolved PL. Streak-camera measurements under non-resonant excitation are performed at the spectral peak of the QW emission, and generally probe the hole lifetimes in 2DEG samples because electrons are always available for recombination (Livescu et al., 1988); in insulating samples the recombination lifetime reflects both carrier lifetimes. We find that the hole lifetimes in the 2DEG samples are <50 ps, and appear to correspond to the initial oscillation damping of 15-50 ps seen in these systems. These data suggest that electron-hole spin exchange, known to be important in intrinsic and p-doped systems (Damen et al., 1991), contributes to the initial electronic spin relaxation seen in the 2DEG samples. Similar measurements on the insulating system reveal a PL lifetime of ~100 ps which is then a lower limit for the hole lifetime and is much longer than their measured transverse spin lifetimes of 5-15 ps. In this insulating sample we expect the presence of excitons to strengthen the electron-hole spin scattering via spin exchange, thus completely relaxing the electronic spins well within their lifetime. In contrast, the 2DEG samples have a shorter hole lifetime and a weaker electron-hole exchange, so that spin relaxation is incomplete and a remanent spin polarization precesses long after the holes have recombined. Because energy relaxation generally occurs well within the measured T_2^* times, we suspect that the electronic spin polarization, excited above the Fermi energy E_F, eventually relaxes into the 2DEG (Damen et al., 1991) where it precesses with dramatically reduced damping.

Figure 9(a) shows the envelope of the 2 T Kerr rotation for a 2DEG structure at temperatures between 5.7 K and 270 K. While the extent of the initial depolarization seems to increase with temperature, the transverse lifetime for later times is only weakly temperature dependent, decreasing from 3.9 ns to 1.3 ns as the temperature is raised from 5.7 K to 270 K. Thus, we clearly resolve spin precession at room temperature, as seen in Figure 9(b). Furthermore, as the temperature is raised above 200 K, the resonance energy begins to sharply decrease and a new energy resonance in the Kerr rotation appears with nearly the same g-factor but at an energy just above the room-temperature ZnSe bandgap. Figure 9(c) shows the Kerr rotation at this energy, for which T_2^* is 0.76ns at room-temperature and B=1 T. We interpret this secondary resonance as arising from the equilibrium transfer of electrons from the 2DEG (producing a decrease in the Kerr resonance energy) into the ZnSe barriers at elevated temperatures. Note that a depletion of the 2DEG obviously complicates any detailed analysis of T_2^* within the QW. In particular, one might expect that the inherent temperature dependence of T_2^* in the QW may be even weaker than observed if one accounts for the depletion of the 2DEG .

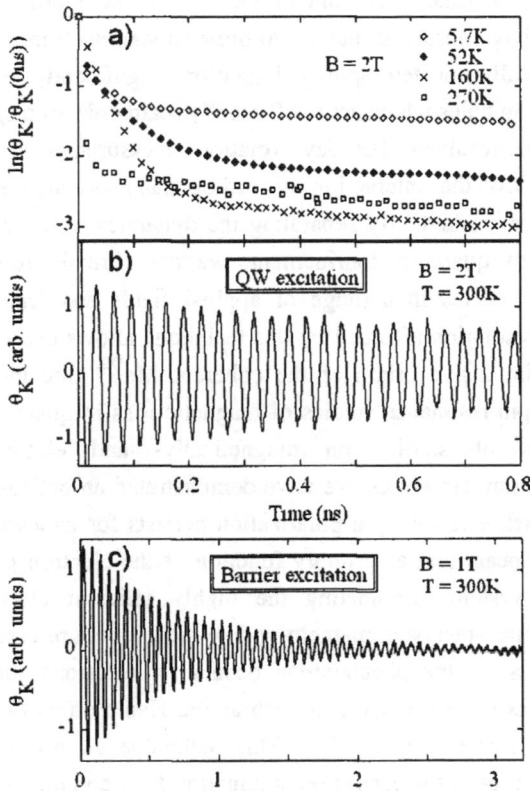

Figure 9. (a) Envelopes of the measured Kerr rotation at B=2T in the sample of Fig. 8 taken with 10^{11} electrons cm^{-2}/pulse. The resonant excitation energy decreases from 2.70 eV to 2.61 eV as the temperature varies from 5.7K to 270K. Amplitude changes have been normalized at zero delay. (b) Room temperature Kerr rotation at 2T at E_{ex}=2.60 eV and 10^{11} electrons cm^{-2}/pulse. (c) Room temperature Kerr rotation at 1T with 5×10^{11} electrons cm^{-2}/pulse. The excitation is at E_{ex}=2.72eV, 20 meV above the ZnSe bandgap energy.

1.3.2 Thin films and bulk semiconductors: II-VI and III-V systems

Systematic studies on doped ZnSe epilayers show spin lifetimes of 1.6 ns at 275K and equilibrium electron concentrations as low as 5×10^{16} cm^{-3}, suggesting that the room temperature spin precession which we observe is a general feature of n doped semiconductors. In fact, these bulk samples exhibit an even weaker temperature dependence than seen in the QW structures, with the spin lifetime decreasing only

20% from 5 K to 300 K. Interestingly, for undoped samples we find elevating the temperature *increases* T_2^* more than tenfold above its low temperature value, perhaps due to the thermal excitation of carriers or the thermal unbinding of excitons. Thus we have observed that a two-dimensional electron gas can act as a reservoir for optically-injected spin polarization, signficantly extending spin lifetimes by removing holes which act to efficiently scatter electron spins.

Thus, time-resolved Faraday rotation measurements in transverse magnetic fields allow the interaction of orthogonally oriented electronic and magnetic spins to be tracked. By separating the dynamics of electrons and holes using the effects of quantum confinement, we can directly resolve the spin relaxation of each species in a range of applied fields and temperatures. The femtosecond time resolution afforded by this technique allows one to recover both the direction and impulsive origin of the coherent Mn^{2+} free induction decay, facilitating direct spin resonance studies of magnetic ions in quantum geometries. Through a series of studies on magnetically- and electronically-doped semiconductor quantum structures, we have demonstrated an optically active solid state system in which electron spin polarization persists for nanoseconds at room temperature. It appears that a primary function of the electron gas is to sweep holes out of the system, terminating the highly efficient electron-hole spin scattering. These time-resolved magneto-optical techniques are essential tools for viewing this process, as the phenomenon occurs for electrons above E_F and is invisible to measures of spin relaxation such as the Hanle effect or time-resolved PL which probe electrons near k=0. This technique allows us witness spin lifetimes which far exceed the carrier recombination time and draws an interesting contrast to systems in which carrier recombination depletes the spin polarization an order of magnitude faster than spin relaxation processes. Furthermore, we expect the behavior described here to occur in other semiconductor materials as well. In bulk, doped GaAs crystals, the benefit is even more dramatic and spin lifetimes become so long that they may be used to establish a resonance between electron Larmor precession and the periodic excitation provided by a pulsed Ti:sapphire laser cavity (Kikkawa and Awschalom, 1998). Here, spins are repeatedly excited at a fixed time interval that is shorter than the lifetime of spin precession. By adjusting the magnetic field so that successive spin injections occur in phase with electron Larmor precession, a resonance may be established in which spin polarization from different laser pulses adds constructively. This resonance condition is met when spins precess an integer number revolutions between laser pulses, and is characterized by an increase in total electron spin at evenly spaced intervals of the applied magnetic field. Figure 10 shows how spin precession may be brought into and out of resonance by varying the applied magnetic field, yielding sharp features whose Lorentzian linewidth yields a spin lifetime exceeding 100 nanoseconds at low temperature. Note that the resonance height decreases with magnetic field, reflecting a field dependence of T_2^*.

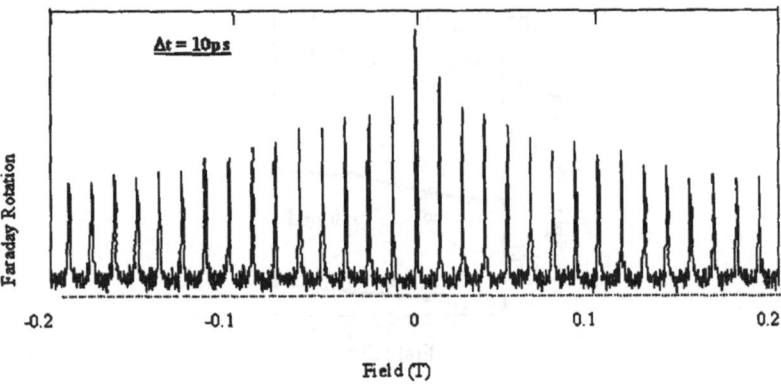

Figure 10. Resonances in Faraday rotation observed in bulk n-type GaAs, arising from the constructive interference of successive spin injections. Data taken at T=4K in a sample having an electrical doping of $n=10^{16}$ cm^{-3}.

Figure 11 shows that in bulk GaAs, lowering the doping level (and thus the electron kinetic energy) uncovers a completely different regime in which spin lifetimes become strongly field- and temperature-dependent and are further extended two orders of magnitude at low temperature. In all of these instances, while the observed precession reveals a memory of the initial spin orientation within the electronic system, its relationship to individual spin coherence is not clear. Though electron-electron interactions of the form $s_i \cdot s_j$ can act to destroy the coherence of individual spins with their initial orientation established by the optical field, they would have no impact on the measured Kerr signal because they do not alter the equations of motion for total electronic spin. These "hidden" decoherences rely on the absence of any spatial dependence to the spin interaction which would then couple to the orbital degrees of freedom, permitting spin relaxation. Because we cannot rule out such processes, only a direct interference experiment can reveal the duration of quantum coherence within the spin precession (Heberle et al., 1995). What we demonstrate here is that the contribution to spin decoherence from all other sources is *quite small, even at* room-temperature.   If these systems do prove to be quantum coherent for nanoseconds at room-temperature, the development of ultra-fast optical techniques to coherently manipulate the spin system would offer new technological opportunities for magnetoelectronics and computation.

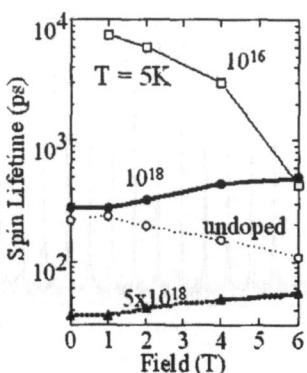

Figure 11. Transverse spin relaxation time T_2^* vs. field measured in commercial wafers of GaAs at T = 5 K with optically-excited carrier densities of N_{ex} = $2*10^{14}$, $2*10^{14}$, $1.4*10^{15}$ and $3*10^{15}$ cm^{-3} for n-type doping levels of n = 0, 10^{16}, 10^{18} and $5x10^{18}$ cm^{-3}, respectively.

As a first step towards manipulating spin coherence in these systems, the above resonance technique was recently modified to demonstrate macroscopic displacement of spin precession.   Since resonant spin amplification occurs via spin accumulation, by applying a lateral electric field to the system that drags the electrons away from the laser excitation point after one excitation, the resonances are destroyed.   This spin "packet" may be imaged by physically scanning the optical probe beam across the sample, and by sweeping the magnetic field at each location. The exact position of the spins can be identified by their age: older spins rotate further in the magnetic field than younger spins do. Using this chronological tagging, overlapping spin packets from distinct spin injections (laser pulses) can be well separated (Kikkawa and Awschalom, 1999). Figure 12 shows the travel of ten consecutive spin injections across distances exceeding 100 microns.   Along the way, environmental spin decoherence is seen to increase only moderately. Combining these developments with the ability to electrically transport spin through quantum structures provides exciting opportunities for coherent spin transport and future technologies.

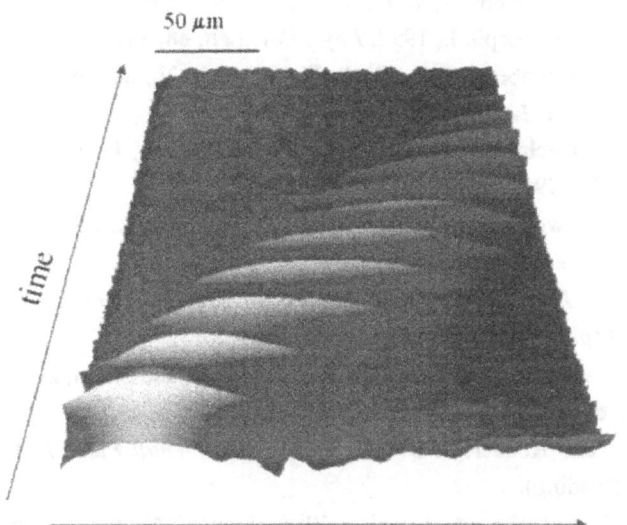

Figure 12. Electrically-driven spin motion in n-type GaAs having $n=10^{16}$ cm^{-3} and E=-37 V/cm. The images reveal a distinct series of coherent spin packets injected every 13 ns using the duty cycle of the laser system. Spin transport is observable at macroscopic displacements and due to free electrons.

1.4 ACKNOWLEDGEMENTS

We thank S. A. Crooker, J.M. Kikkawa, F. Flack, and I.P. Smorchkova for their contributions to this research. In addition, we acknowledge financial support from the ARO DAAG55-98-1-0366 , the ONR N00014-99-1-0077 and N00014-99-1-0071, the NSF DMR 97-01072 and NSF DMR 97-01484, the AFOSR F49620-99-1-0033, and the NSF Science and Technology Center for Quantized Electronic Structures (DMR 91-20007).

1.5 REFERENCES

Awschalom, D. D. and Samarth, N., 1993, in *Optics of Semiconductor Nanostructures*, edited by F. Hennenberger, S. Schmitt-Rink, and E.O. Gobel (Akademie Verlag, Berlin). pp. 291-311.

Awschalom, D.D. and Samarth, N., 1998, in *Dynamical Properties of Unconventional Magnetic Systems*, NATO ASI Series Volume 349, eds. A. T. Skjeltorp and D. Sherrington [Kluwer Academic, The Netherlands, 1998].

Bar-Ad, S. and Bar-Joseph, I., 1992, *Phys. Rev. Lett.* **68,** 349

Crooker, S. A., Baumberg, J. J., Flack, F., Samarth, N., and Awschalom, D. D., 1996, *Phys. Rev. Lett.* **77,** 2814.

Crooker, S. A., Tulchinsky, D. A., Levy, J., Awschalom, D. D., Garcia, R., and Samarth, N., 1995, *Phys. Rev. Lett.* **75,** 505.

Crooker, S. A., Awschalom, D. D., Baumberg, J. J., Flack, F., and Samarth, N., 1997, *Phys. Rev. B* **56** , 7574.

Crooker, S. A., Awschalom, D. D., and Samarth, N., 1995, *IEEE Journal of Selected Topics in Quantum Electronics* **1**, 1082.

Damen, T. C., Vina, L., Cunningham, J. E., Shah, J., and Sham, L. J., 1991, *Phys. Rev. Lett.* **67**, 3432.

Fukushima, E. and Roeder, S. B. W., 1981, *Experimental Pulse NMR*, (Addison-Wesley, Reading).

Furdyna, J. K. and Kossut, J., eds., 1988, *Diluted Magnetic Semiconductors*, (Academic, San Diego).

Gershenfeld, N. A. and Chuang, I. L., 1997, *Science* **275**, 350.

Gider, S. et al.., 1996, *Appl. Phys. Lett.* **69**, 3269.

Heberle, A. P., et al., 1994, *Phys. Rev. Lett* **72**, 3887

Heberle, A. P., Baumberg,J. J., and Kohler, K., 1995, *Phys. Rev. Lett.* **75**, 2598.

Kavokin, K. V. and Merkulov, I. A., 1997, *Phys. Rev. B* **55**, 7371.

Kikkawa, J. M., Smorchkova, I. P., Samarth, N., and Awschalom, D. D., 1997, *Science* **277** , 1284.

Kikkawa, J. M. and Awschalom, D. D., 1998, *Phys. Rev. Lett.* **80**, 4313.

Kikkawa, J. M. and Awschalom, D. D., 1999, *Nature* **397**, 139.

Kuhn-Heinrich, B. and Ossau, W., 1995, in *II-VI Compunds and Semimagnetic Semiconductors*, edited by H. Heinrich and J.B. Mullin (Trans. Tech. Publications, Switzerland).

Livescu, G., et al., 1988, *IEEE J. Quantum Electron.* **24**, 1677.

Martin, R. W., et al., 1990, *Phys. Rev. B* **42**, 9237

Prinz, G., 1990, *Science* **250**, 1092.

Prinz, G., 1995, *Physics Today* **48**, 58.

Samarth, N. and Furdyna, J. K., 1988, *Phys. Rev. B.* **37**, 9227.

Stühler, J. *et al.*, 1995, *Phys. Rev. Lett.* **74**, 2567.

Tornow, M. et al., (1996), *Phys. Rev. Lett.* **77**, 147.

Twardowski, A., et al., 1983, *Sol. Stat. Commun.* **48**, 845.
Uenoyama, T. and Sham, L. J., 1990, *Phys. Rev. Lett.* **64**, 3070.

3 Synthesis of Nanowires from Laser Ablation

S. T. Lee, N. Wang, Y. F. Zhang, Y. H. Tang, I. Bello and C. S. Lee
Center of Super-Diamond and Advanced Films, Department of Physics and Materials Science, The City University of Hong Kong, Kowloon, Hong Kong

ABSTRACT

One-dimensional nanoscale materials with uniform size have been synthesised by laser ablation technique. In contrast to the traditional vapour-liquid-solid technique (using metal particle catalysts), oxides have been discovered to play a critical role in enhancing the formation and growth of semiconductor nanowires. A new growth mechanism was therefore proposed based on the microstructure and various morphologies of the nanowires observed.

In this paper, we show that laser ablation technique is an effective method to synthesize ultra long, highly pure, and uniformly sized one-dimensional nanoscale materials in bulk-quantity for fundamental research and promising technological applications. Synthesis of one-dimensional nanostructures has been a challenge in materials science. Since the 1960's, semiconductor whiskers have been synthesized by the vapour-liquid-solid (VLS) reaction (Wagner et al., 1964, Givargizov, 1975). In the recent year, a variety of techniques, such as arc-discharge, chemical-vapour-deposition (CVD), photolithography technique combined with etching and scanning tunnelling microscopy, are employed to synthesize carbon nanotubes, metallic and semiconductor nanowires (Iijima, 1991, Maluf et al., 1992, Namatsu, et al., 1997, Ono, et al., 1997, Hasunuma, et al., 1997). Recently, bulk-quantity semiconductor nanowires have been successfully synthesized by a novel method, namely, laser ablation of metal-containing Si targets (Wang, et al., 1998a, Zhang, et al., 1998, Morales, A.M. and Lieber, 1998, Yu, et al., 1998). We extend this technique to synthesize large-quantity one-dimensional nanostructures of different material systems. For semiconductor nanowire synthesis, we proposed a new mechanism based on oxide-assisted growth of nanowires.

The experimental apparatus simply consists of an evacuated quartz tube (500 Torr, the carrying gas is Ar) placed inside a tube furnace. The optimal temperatures of the substrates depend on the materials used. For Si and Ge nanowires growth, the temperatures are 930-950 °C and 690-700 °C respectively.

The targets are mixtures of highly pure semiconductor and oxide powders, such as Si/SiO_2, Si/Fe_2O_3 or Ge/GeO_2. Details of this method can be found in the references of Wang, et al., 1998a and Zhang, et al., 1998.

As reported by different groups, laser ablation or thermal evaporation of metal-containing Si powder could produce Si nanowires (Wang, et al., 1998a, Zhang, et al., 1998, Morales, A.M. and Lieber, 1998, Yu, et al., 1998). Based on the assumption that metals or metal-silicide nanoparticles act as the critical catalyst during growth, the growth mechanism for Si nanowires has been attributed to the VLS reaction. From our investigations, however, metals were shown to be unnecessary for Si nanowires growth (Wang, et al., 1998b, Wang, et al., 1998c). Si nanowires of limited quantity were obtained even when pure Si powder was used as the target. Since Si oxide was inevitably present in the targets, it was then suggested that Si oxide might have a special effect on the nanowire growth. As expected, Si nanowire growth was indeed greatly enhanced when SiO_2-containing Si powder targets were ablated. As shown in Fig.1, the yield of Si nanowire product obtained from SiO_2-containing targets was always greater than, and up to 30 times (at 50%wt SiO_2), the product generated from metal-containing targets (0.03 mg/hr). Even by adding 90% SiO_2 into the target, the yield was still higher than that from a metal-containing target. In fact, Si nanowires were produced (0.02 mg/hr) even by using the target containing 99% SiO_2. However, pure SiO_2 target did not produce Si nanowires.

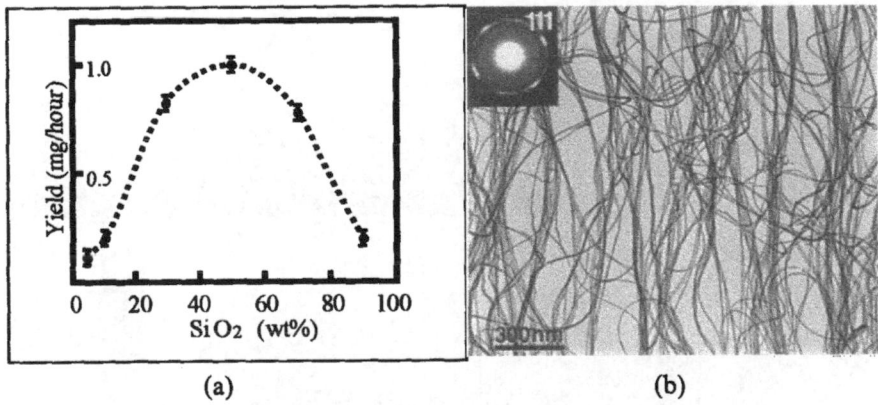

(a)   (b)

Figure 1 (a) Yields of Si nanowire product obtained by using the targets containing different percentages of SiO_2. (b) TEM image shows the morphology of Si nanowires grown from the SiO_2-containing target.

From the following experiment using two targets, we found that SiO_2 was constantly needed for nanowire nucleation and growth. A SiO_2-containing target was ablated first in to form Si nanowire nuclei. Subsequently, pure Si target was ablated to induce further growth. However, no detectable growth of Si nanowires was observed. This is contrary to the VLS growth process in which metal particles, once deposited initially to form the mediating solvent, will continue to *catalyse Si* nanowire growth via the molten eutectic metal-Si alloy present at the tips of Si nanowires.

Using SiO_2-containing targets, we grew Si nanowires at a very high yield. The thickness of the nanowire product was approximately 1 cm (3-hour ablation). Highly pure Si nanowires grew nominally in the same direction with a uniform diameter (Fig.1(b)). Selected-area electron diffraction (SAED) (inset in Fig.1(b)) showed obviously that the intensity of Si (111) diffraction ring exhibited strong texture. This indicated that Si nanowires had a similar growth direction which was consistent with our previous observation (Wang, et al., 1998a) that the axes of Si nanowires were generally along the <112> direction. Figs.2 (a)-(d) show the formation of the nanowire nuclei during the initial growth stage. Si nanoparticles were precipitated in the initially deposited SiO_x matrix as identified by SAED and electron energy dispersive spectrometer (EDS). Some nanoparticles piled up on the SiO_x surface. The favorable particles (or nuclei of nanowires) stood alone and underwent faster growth because their growth direction was normal to or away from the substrate surface which is the preferred growth direction. No detectable metal catalyst or impurity formed at the tips of the nuclei. Each nucleus simply consisted of a Si crystalline core and an amorphous outer layer. The EDS analysis showed that the amorphous outer layer was silicon oxide. The Si crystal core at nanowire tip contained a high density of defects (Fig. 2 (a)). Most of the defects were quite similar to the planar defects of stacking faults and micro-twins observed in our previous study (Wang, et al., 1998a).

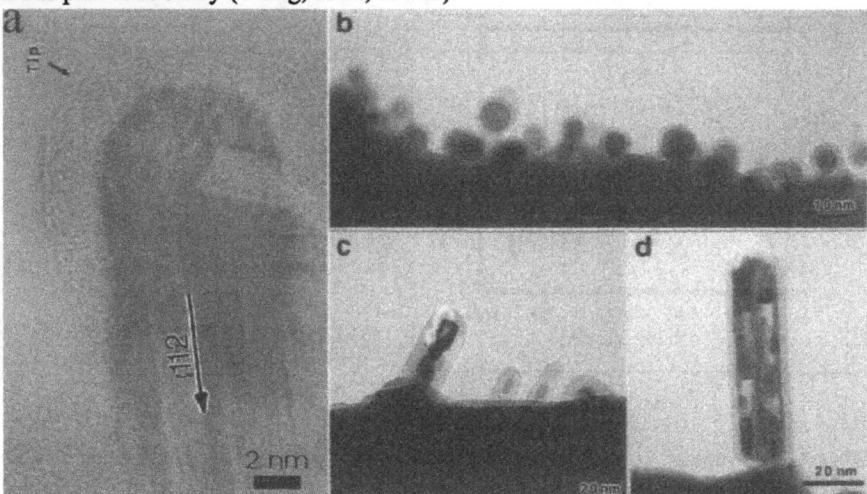

Figure 2 (a) High-resolution TEM image showing the microstructure of Si nanowire tip.
(b)-(d) Nucleation stages of Si nanowires

Based on the present results and our observations (Wang, et al., 1998c) that the vapour phase generated from Si/SiO_2 target was Si_xO_y (x =0.51-0.60, y = 0.49-0.40), we propose that the growth mechanism is silicon oxide assisted. Fig.3 shows this mechanism schematically. The vapor phase of Si_mO (m>1) generated by laser ablation is the key factor. The nucleation of nanoparticles is assumed to form at the substrate by the following decompositions of Si oxide:

$$Si_m O \rightarrow Si_{m-1} + SiO \ (m > 1) \tag{1.1}$$

and

$$2SiO \rightarrow Si + SiO2 \tag{1.2}$$

These decompositions result in the precipitation of silicon nanoparticles or the nuclei of Si nanowires, which have been confirmed by our TEM observations. Si nanowire growth may be determined by the following factors: (i) the catalytic effect of the $Si_m O$ layer formed at nanowire tips. Since the surface melting temperature of nanoparticles can be much lower than that of their bulk materials, the materials at Si nanowire tips (similar to the case of nanoparticle) may be in/or near the molten states. The presence of molten states can enhance atomic absorption, diffusion and deposition; (ii) the SiO_2 component in the shells, which is formed from $Si_m O$ decomposition retards the lateral growth of nanowires; (iii) the main defects in Si nanowires are stacking faults along the nanowire growth direction of <112> (normally containing 1/6[211], easily moving, and 1/3[111], not moving partial dislocations) and microtwins. The presence of these defects at the tip areas should result in the fast growth of Si nanowires since dislocations are known to play a very important role in crystal growth; (iv) the {111} surface, which has the lowest surface energy among the Si surfaces, played an important role for Si nanowire nucleation and growth. Since surface energy becomes more important when the crystal size is in nanometer scale, the presence of the {111} surfaces parallel to the axes of the nanowires reduces the system energy. The last two factors determine that only those nuclei with their <112> directions parallel to the preferred growth direction can grow fast (see Fig.3). For those non-favorable nuclei, growth should terminate and re-nucleation occurs. This may result in the formation of kinks in nanowires or further re-nucleation if the newly formed nucleus is not favorable. Re-nucleation will result in the formation of Si nanoparticle chains (Wang, et al., 1998c).

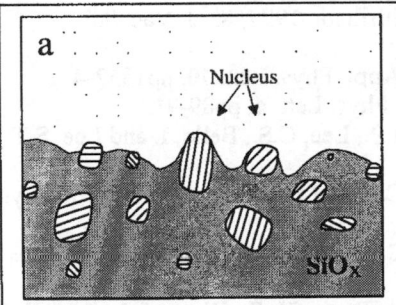 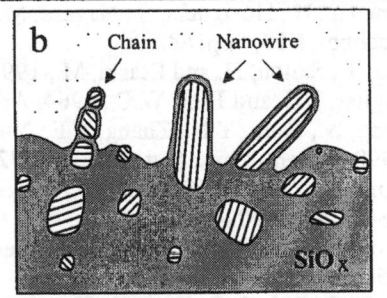

Figure 3 Nucleation and growth mechanisms for Si nanowires. The parallel lines indicate the <112> orientation. (a) Si oxide vapour phase deposits and forms the matrix first. Then Si nanoparticles precipitate. (b) Preferred oriented nanoparticles grow fast and form nanowires. Non-preferred oriented nanoparticles may form nanoparticle chains.

The oxide-assisted Si nanowire growth mechanism was further supported by the experiment that Fe_2O_3-containing Si target also produced a high yield of Si nanowire product (~0.3 mg/hour). The reaction was similar to the case of SiO_2-containg targets. Si oxide vapor phase was generated and therefore enhanced the nanowire growth. In addition, we have succeeded in using GeO_2-containing targets to synthesize Ge nanowires. TEM studies show that the morphology of Ge nanowires is similar to that of Si nanowires, i.e. each Ge nanowire consists of a crystalline Ge core and a thick amorphous oxide shell. This is not surprising since Ge and Si have similar structure and properties. Our investigations revealed no significant difference in the morphology or microstructure of Si or Ge nanowires synthesized from metal-, SiO_2- or Fe_2O_3- containing targets. Thick oxide shells were observed in all case. The common feature of these kinds of powder targets is the presence of oxide. This reaffirms that oxide plays a special role in the nucleation and growth of the high quality nanowires. For other materials systems, we have successfully synthesized the nanowires of SiO_2, amorphous carbon and SiC by using the same experiment configuration (Zhou, et al, 1998). The large quantity of nanowires synthesized by this novel technique offer exciting opportunities for fundamental research and technological applications.

Acknowledgements

This work was supported by Strategic Research Grant of the City University of Hong Kong and Research Grant Council of Hong Kong.

Reference:

Givargizov, 1975, E.I., J. Cryst. Growth , **32**, pp20-30.
Hasunuma, R., Komeda, T., Mukaida, H., Tokumoto, H., 1997, J. Vac. Sci. Technol. **B15**, pp1437-41.
Iijima, S., 1991, Nature **354**, pp56-58.
Maluf, N.I. and Pease, R.F.W., 1992, J. Vac. Sci. Technol. **B9**, pp2986-91.
Morales, A.M. and Lieber, C.M., 1998, Science, **279**, pp208-11.
Namatsu, H., Horiguchi, S., Nagase, M., Kurihara, 1997, K., J. Vac. Sci. Technol. **B15**, pp1688-96 .
Ono, T., Saitoh, H. and Esashi, M., 1997, Appl. Phys. Lett. **70**, pp1852-4.
Wagner, R.S. and Ellis, W.C., 1964, Appl. Phys. Lett. **4**, pp89-91.
Wang, N., Tang, Y.H., Zhang, Y.F., Yu, D.P., Lee, C.S., Bello, I. and Lee, S.T., 1998a, Chem. Phys. Lett., **283**, 368-72.
Wang, N., Zhang, Y.F., Tang, Y.H., Lee, C.S. and Lee, S.T., 1998b, Appl. Phys. Lett., **73**, pp3902-4.
Wang, N., Tang, Y.H., Zhang, Y.F., Lee, C.S. and Lee, S.T., 1998c, Phys.Rev. **B58**, ppR16024-26.
Yu, D. P., Bai, Z. G., Ding,Y., Hang, Q. L., Zhang, H. Z., Wang, J. J., Zou, Y. H., Qian, W., Xiong, G. C., Zhou, H. T. and Feng, S. Q., 1998, Appl. Phys. Lett. **72**, pp3458- 60.
Zhang, Y.F., Tang, Y.H., Wang, N., Yu, D.P., Lee, C.S., Bello, I. and Lee, S.T., 1998, Appl. Phys. Lett. , **72**, pp1835-37.
Zhou, S.T., et al., 1998, not published.

4 Level Spacings in Small Clusters

Ping Sheng and Lin Yi
Department of Physics, Hong Kong University of Science and Technology, Clear Water Bay, Kowloon, Hong Kong, China

ABSTRACT

By taking into account the Coulomb and Breit interactions between the electrons, we carried out self-consistent Hartree-Fock calculations of the energy levels in randomly-shaped jellium clusters with up to 276 electrons. The level spacing statistics is found to differ qualitatively from the Wigner distribution and better correspond with the Poisson distribution. Prediction of significantly enhanced far-infrared absorption is in excellent accord with experimental observations. Breit interaction is identified as key to the semi-metallic behavior persistent in small metal clusters at energy scales much lower than the average level spacing.

INTRODUCTION

In metallic clusters, quantization of motion causes the electronic energy levels to be discrete, with a mean level separation which increases with decreasing cluster size (Frohlich, 1937). Thus it is the conventional wisdom that for small metallic clusters, there can be a metal-insulator transition when the excitation energy or kT, where k is the Boltzmann constant and T the temperature, falls below the mean level spacing (Kubo, 1962). In particular, the far-infrared absorption was expected to be small, since there should be a dearth of level spacings for the accompanying electronic transitions. In 1976, a Cornell experiment (Granqvist et al., 1976) showed that, contrary to the usual expectation, the far-infrared absorption was orders of magnitude larger than that predicted theoretically. During the subsequent years, additional experiments have shown that while some of the "anomalous" absorption is due to cluster-cluster aggregation and some due to absorbing surface coatings, there is nevertheless a large discrepancy yet to be accounted for (Tanner et al., 1990).

In this work, we show that the usually ignored Breit interaction between the electrons (Breit, 1930) could be the explanation for this long-standing puzzle. The inclusion of Breit and Coulomb interactions in our calculations is also found to have significant implications for the level spacing distribution in small metallic clusters. Instead of the usually assumed Wigner distribution

$P(\sigma) = (\pi\sigma/2)\exp(-\pi\sigma^2/4)$ (Mehta, 1991), where σ denotes the level spacing normalized by its mean, we find the level spacing distribution to be better described by the Poisson distribution, $P(\sigma) = \exp(-\sigma)$, when the electron-electron interactions are accounted for. In contrast to the Wigner distribution, which exhibits a quasigap at small values of σ, the Poisson distribution has a predominance of small level spacings.   Therefore instead of the expected insulator behavior, e.g., small far-infrared absorption, small metallic clusters should be inherently semi-metallic in character.   Indeed, our calculated far-infrared absorption is orders of magnitude larger than the previous predictions, in good agreement with the experiments (Yi and Sheng, 1998).

THE ELECTRON-ELECTRON INTERACTIONS

The starting point of our calculations is the N-electron Dirac-Breit (DB) Hamiltonian (Bethe and Salpeter., 1957):

$$H = \sum_{i=1}^{N}(c\vec{\alpha}_i \cdot \vec{p}_i + \beta_i mc^2 + V_o) + \sum_{j<i}^{N} \frac{e^2}{|\vec{r}_i - \vec{r}_j|} -$$

$$\sum_{j<i}^{N} \frac{e^2}{2|\vec{r}_i - \vec{r}_j|}[\vec{\alpha}_i \cdot \vec{\alpha}_j + \frac{(\vec{\alpha}_i \cdot \vec{r}_{ij})(\vec{\alpha}_j \cdot \vec{r}_{ij})}{|\vec{r}_i - \vec{r}_j|^2}], \tag{1}$$

where the first summation term represents the Dirac Hamiltonian for N electrons, c the speed of light, m the electron mass, $\vec{p}_i = -i\hbar\vec{\nabla}_i$, $\vec{\alpha}_i$ and β_i are the Dirac matrices of the i th electron, and V_o is the jellium potential, defined to be $-(E_F + W)$ inside the cluster boundary, with W deonting the work function, and zero otherwise. The second term of (1) gives the Coulomb interaction between the electrons, with e denoting the electronic charge, and *the third term is the so-called Breit operator*, which is the source of the spin-spin interaction. By minimizing the expectation value of the DB Hamiltonian, using as the basis the linear combinations of antisymmetrized Slater determinants product wavefunctions, and keeping terms only to the quadratic order of the expectation value of $\vec{p}/mc$, a non-relativistic Breit-Hartree-Fock (BHF) equation has been obtained (Yi and Sheng, 1998). By expanding the eigenfuctions in appropriate basis functions, the BHF equation may be written in the form of an iterative matrix equation for the expansion coefficients $\{C\}$, whose self-consistent solution gives all the energy eigenvalues of the system.

LEVEL SPACING STATISTICS AND ELECTRONIC SUSCEPTIBILITY

By using parameter values appropriate to bulk Al, i.e., $E_F = 11.7$ eV,
$W = 4.25$ eV (Ashcroft and Mermin, 1976), we have carried out numerical
calculations for randomly-shaped clusters with 276 electrons, or about 1 nm in
diameter. The left panel of Fig. 1 is a plot of the integrated level spacing data
from *four* randomly-shaped clusters as a function of σ, with the corresponding
distributions shown in the insets. For comparison, we also show the curve
representing the integral of the Wigner surmise. The effect of adding interactions
is clearly *seen to be increasing departure from the Wigner distribution*. In Fig.
1(c), the data clearly disagree with the Wigner surmise and are better described
by the Poisson distribution, the integral of which is shown by the dotted line.
This Poisson-like distribution is recognized to be due to the existence and overlap
of three energy scales: shape quantization, Coulomb interaction, and Breit
interaction. Our result implies an abundance of level spacings smaller than the
average value.

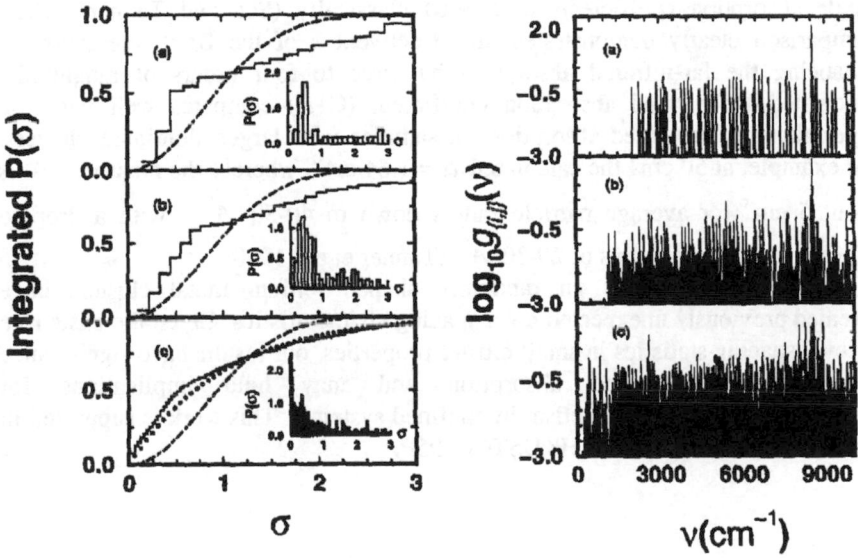

Figure 1 Left panel: Integrated level spacing distribution plotted as a function of
the normalized level spacing variable σ. The associated distributions are shown
as insets. The integral of the Wigner distribution is denoted by the dash-dotted
line. The integral of the Poisson distribution, shown only in (c), is given by the
dotted line. (a) Noninteracting electrons, (b) Coulomb interaction only, (c)
Coulomb plus Breit interactions. Right panel: Associated oscillator strength
$g_{\{i,j\}}$.

In the right panel of Fig. 1 we plot the calculated, symmetrized oscillator strength $g_{ij} + g_{ji} = g_{\{i,j\}}$ as a function of the optical transition frequency v for the three corresponding cases shown in the left panel. Here the oscillator strength g_{ij} is defined as in Yi and Sheng (1998). The results of the four different shaped clusters are plotted together. For the noninteracting case and the Coulomb interaction (only) case, there are clear gaps (>300 cm^{-1}), indicating an insulator-like behavior at low frequencies.

FAR-INFRARED ABSORPTION

Optical absorption can be directly calculated from the oscillator strengths, as described in Yi and Sheng (1998). We wish to compare our calculated results with the far-infrared absorption experiments in which the metal clusters are dispersed in a KCl matrix (Tanner et al., 1990). Shown in Fig. 2 are the magnitude and frequency dependencies of the absorption coefficient calculated for the three cases, with $x = 0.8\%$. Here x denotes the volume fraction of metal clusters embedded in KCl. Also shown is the sum of the electric and magnetic dipole absorption coefficients calculated classically (Sen and Tanner, 1982). Comparison clearly demonstrates the effectiveness of the Breit interaction in enhancing the far-infrared absorption by three to four orders of magnitude. Calculated far-infrared absorption coefficient (C+B) compares well with the experimentally measured absorption on samples with larger aluminum clusters. For example, at 50 cm^{-1} the calculated $\alpha = 1.64$ cm^{-1}, whereas the measured α is about 5 cm^{-1} for average particle radius down to 40-50 A°, with a drop to 1-2 cm^{-1} for particle radius of 20-30 A° (Tanner et al., 1990).

Our calculations on randomly shaped jellium metal clusters have revealed previously unexpected level spacing characteristics. Given the basic role of level spacing statistics in small cluster properties, our results have significance beyond the far-infrared absorption and may hold implications for superconductivity and magnetism in confined systems. This work is supported in part by the grant RGC95/96.HKUST612/95P.

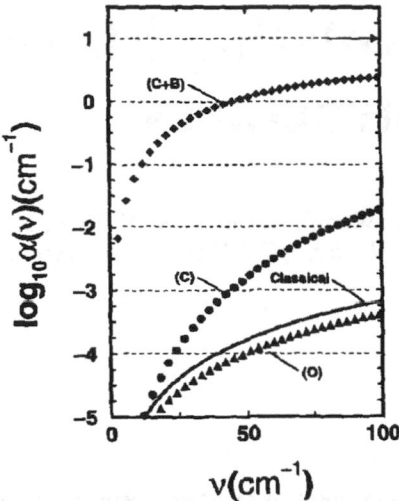

Figure 2 Frequency dependence of the calculated absorption coefficient for a sample of KCl with 0.8 volume-percent of dispersed 276-electron Al clusters. The Al clusters are assumed to have four different random shapes with equal probability. The four cases shown are for noninteracting electrons (solid triangles), the classical prediction (solid line), Coulomb interaction only (solid circles), and Coulomb plus Breit interactions (solid diamonds). The arrow indicates the experimentally measured absorption coefficient at 100 cm^{-1} for samples of 0.8% Al clusters dispersed in KCl (Tanner et al., 1990).

REFERENCES

Ashcroft, N. W. and Mermin, N. D., 1976, *Solid State Physics* (Holt, Rinehart and Winston, New York) p. 364.

Bethe, H. A. and Salpeter, E. E., 1957, *Quantum Mechanics of one-and two-Electron Atoms* (Springer-Verlag, Berlin) pp. 170-178.

Breit, G., 1930, *Physical Review* **36**, pp. 383-397.

Frohlich, H. , 1937, *Physica (Utr.)* **4**, pp. 406-412.

Granqvist, R. A., Buhrman, R. A., and Sievers, A. J., 1976, *Physical Review Letter* **10**, pp. 625-629.

Kubo, R., 1962, *Journal of Physical Society of Japan* **17**, pp. 975-986.

Mehta, M. L., 1991, *Random Matrices* (Academic Press, 2nd edition) pp. 18-20.

Sen, P. N. and Tanner, D. B., 1982, *Physical Review B* **26**, pp. 3582-3587.

Tanner, D. B., Kim, Y. H., and Carr, G. L., 1990, in *Physical Phenomena in Granular Materials*, eds Cody, G. D., Geballe, T. H., and Sheng, P. (Materials Research Society, Pittsburgh) pp. 3-14.

Yi, L. and Sheng, P., 1998, *Physica A*, to appear.

5 A Phenomenological Model of Percolating Magnetic Nanostructures

T. K. Ng[1], S. K. Wong[1], B. Zhao[1], X. N. Jing[1], X. Yan[1] and P. M. Hui[2]
[1]*Department of Physics, HKUST and* [2]*Department of Physics, CUHK*

ABSTRACT

Transport and magnetotransport properties were analysed systematically in percolating nanostructures such as Ni-rich $NiFe - SiO_2$ and $Fe - SiO_2$ films. These granular magnetic films exhibit giant Hall effect. We identified features which are common and unique to these systems. Among the features are the correlation between a $-\log T$ like temperature dependent resistivity and a particle size distribution having a large fraction of small nanometer sized particles, and the power law dependence between the magnetoresistivity and the room temperature resistivity. Assuming the presence of nanometer sized particles in the percolating conduction channels whose contributions are sensitive to temperature and the external magnetic field, we developed a phenomenological model to explain all the common features.

CHARACTERISTIC FEATURES IN GRANULAR FILMS WITH GIANT HALL EFFECT

In a series of experiments (Pakhomov *et al.*, 1995; Jing *et al.*, 1996; Zhao and Yan, 1997), Yan and coworkers studied the physical properties of ferromagnetic granular composite films exhibiting giant Hall effect (GHE). GHE refers to a 10^4-fold enhancement in Hall resistivity observed in $(NiFe)_x - (SiO_2)_{1-x}$ and $Fe_x - (SiO_2)_{1-x}$ films for x near the percolating volume fraction x_c as compared with that of a homogeneous ferromagnetic metal. We identify a few characteristic features unique and common to these systems. A large resistivity with $\rho(T) \sim -\log T$ was observed in both the $NiFe - SiO2$ and $Fe - SiO_2$ films which have carriers of different signs (Zhao and Yan, 1997). The resistivity reduces and the $\log T$ like behaviour disappears for samples annealed at high temperatures, accompanied by the disappearance of GHE. Figure 1 shows $\rho(T)$ (normalized to the value at 5K) on a logarithmic scale for $Fe_{0.53}$-$(SiO_2)_{0.47}$ under different annealing temperatures

T_a. For T_a<300°C, under which GHE exists (ρ_{xys}~ 0.1mΩ-cm, ρ~0.1 Ω-cm), $\rho(T)$~ -log T. For T_a>300°C, under which GHE disappears (ρ_{xys} ~ 0.001 mΩ-cm), $\rho(T)$ shows a metallic behaviour with a considerably lower resistivity (ρ~0.1mΩ-cm). The same qualitative features were also found in *NiFe-SiO₂* films (Jing *et al.*, 1996). Another interesting feature is the variation in the magnetoresistivity as a function of metal volume fraction in the range of 52-65% for the as-deposited films. Figure 2 shows the magnetoresistivity -$\Delta\rho$ versus resistivity on a log-log scale for *NiFe-SiO₂* (circle) and *Fe-SiO₂* (square) samples measured at 300K. The values of -$\Delta\rho$ in *Fe-SiO₂* have been divided by 4 to take into account the difference in the saturation magnetization because -$\Delta\rho$ scales with M^2 for a spin-dependent process. The solid line is the fitting according to -$\Delta\rho$~ρ^b with b~ 1.2, showing that within experimental error the *same power law* is found for both systems, suggesting that the same spin-dependent transport mechanism is functioning regardless the sign of the carriers and the details of the microstructures.

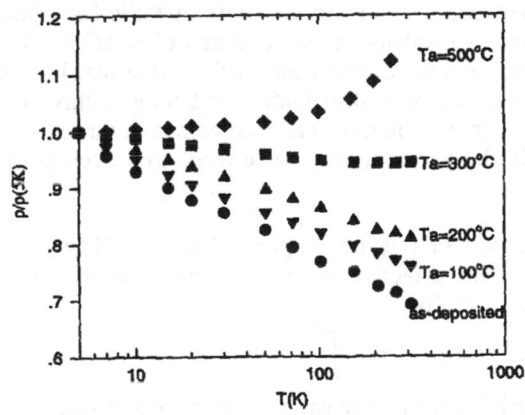

Fig.1 Resistivity as a function of temperature for $Fe_{0.53}$- $(SiO_2)_{0.47}$ films with various annealing conditions.

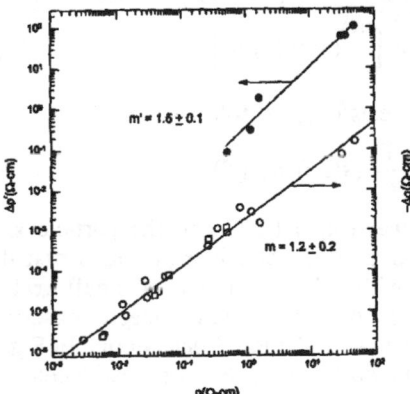

Fig.2 Magnetoresistivity, -$\Delta\rho$, as a function of resistivity ρ for *NiFe – SiO₂* (open circle) and *Fe – SiO₂* (open square), and temperature dependent part of resistivity, $\Delta\rho'$, versus ρ for *NiFe – SiO₂* (filled circle).

PHENOMENOLOGICAL MODEL

In what follows, we introduce a phenomenological model capable of explaining the common features. Our physical picture is based on the TEM observation that even in the metallic phase, well defined metallic clusters percolating throughout the whole sample are absent. Instead the conducting channels appear to be composed of metallic granular particles with a wide distribution in size (Jing et al., 1996). The average distance between metallic grains decreases as the metal volume fraction increases. Our model assumes that the transport properties can be described by a continuum percolation model. Near the percolation threshold, $\rho(T,H) = \rho_0(x_{eff}(T,H) - x_c)^{-t}$, where ρ_0 is a material dependent parameter, t is a critical exponent, $x_{eff}(T,H)$ is the effective conducting volume fraction which depends on the temperature and the external magnetic field. The physical picture is that we have a collection of small particles with the conduction from one to another constitutes the percolating path. However, a metallic particle can be considered as conducting only if there is no energy barrier against the transport of electrons in and out of the particle. The temperature dependence comes in through the thermal energy kT. If the thermal energy of electrons is larger than the energy level spacing inside the particle, $\Delta E(s) \sim h^2/8\pi^2 m^* s^2$, where m^* is the effective mass and s is the linear dimension of the particle, or $\Delta E(s) \sim e^2/\varepsilon s$, where ε is an effective dielectric constant, the particle may be considered to be conducting. In the experimental systems, we have a large number of particles with $s < 3nm$. The important point is that, regardless the origin of the s-dependence, small particles with metallic behaviour at high temperatures would behave as insulator at temperatures $kT \le \Delta E(s)$. The continuum percolation model with a network of resistors with a wide distribution of resistance depending on the particle size distribution and temperature thus captures the basic features. The magnetic field dependence comes in through the alignment of the magnetization in the particles and hence leads to a decrease in resistance.

In continuum percolation, the exponent t is no longer universal. Here, it is simply taken as a parameter. The effective volume fraction in vanishing magnetic field at temperature T is given by

$$x_{eff}(T,0) \sim \int_{s(T)}^{\infty} (s^3)n(s)ds = x_0 - \int_0^{s(T)} (s^3)n(s)ds, \tag{1}$$

where $n(s)ds$ is the number of particles with dimensions between s and $s+ds$, and x_0 is the metal volume fraction. Only particles with $s>s(T)$ contribute to conduction with $s(T)$ determined by $kT = \Delta E(s)$. Using Eq.(1), we have

$$\rho(T,0) \sim \rho_\infty \left(1 + t(\frac{\rho_\infty}{\rho_0})^{\frac{1}{t}} \int_0^{s(T)} (s^3)n(s)ds \right), \tag{2}$$

where $\rho_\infty = \rho_0(x-x_c)^{-t}$ is the high temperature resistivity. It follows that

$$\frac{d\rho(T,0)}{dT} = t\rho_\infty (\frac{\rho_\infty}{\rho_0})^{\frac{1}{t}} \frac{ds(T)}{dT}(s(T))^3 n(s(T)), \tag{3}$$

which provides a precise relationship between $d\rho(T,0)/dT$ and the particle size distribution $n(s(T))$. From Eq.(2), $\rho(T)$ has a log T behaviour provided that the distribution $n(s) \sim s^{-4}$, i.e., a distribution with a large number of small grains. Note that rapid increase in $n(s)$ as s decreases was indeed observed experimentally (Jing et al., 1996) for samples with the log T behaviour. In Fig.3, we show the size distribution derived from Eq.(3) using temperature dependent resistivity data as input and compare it with experimental data obtained by TEM. For both annealed and as deposited films, which have qualitatively different features as shown in Fig.1, the theory gives distributions in good agreement with experimental data, indiciating that our model does capture the essential physics.

Our model also predicts a scaling relationship between the saturated magnetoresistivity and the resistivity in the form $\Delta\rho = -t\rho_\infty(\rho_\infty/\rho_0)^{1/t}\Delta x_M(T)$, where $\Delta x_M(T) = x_{eff}(T, H>H_s) - x_{eff}(T,0)$ is the difference between the effective metallic volume fractions at high field (higher than saturation field) and zero field. Using $300K$ as the high temperature limit, we have $\Delta\rho \sim \rho^b$ with $b = 1+t^{-1}$, which, within experimental accuracy, is consistent with experimental values of $b \sim 1.2 \pm 0.1$ and $t \sim 3.6 \pm 1.0$ (Pakhomov *et al.* 1995). The theory also predicts (see Eq.(2)) that $\Delta\rho' \equiv \rho(T,0) - \rho_\infty \sim \rho_\infty^{1+1/t}$. Taking $T=77K$, Fig.2 shows $\Delta\rho'$ as a function of $\rho(300K)$ for $(NiFe)_x - (SiO_2)_{1-x}$ samples and a scaling relationship exists, as predicted by our model.

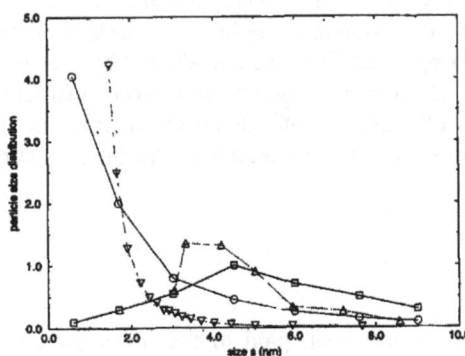

Fig.3 Particle size distributions ($\times$ 10^{11}/mm³) determined by TEM (solid line) and from theory (dash line) for (a) $Ni_{0.55} - (SiO_2)_{0.45}$ as deposited film (O (TEM), ∇(theory)), and (b) $Ni_{0.55} - (SiO_2)_{0.45}$ annealed at 520°C ((TEM), Δ(theory)).

In summary, we proposed a phenomenological model of magnetic granular films near the percolation threshold which is capable of explaining the common and unique features in these systems. In addition to the features discussed, the model also predicts a scaling relationship between the Hall number and the resistivity which is consistent with experimental data (Wong *et al.*, 1998).

We would like to thank P. Sheng, K.K. Fung, Y. Xu, A. Pakhomov, J. Xhie, S. Leung, and L.T. Wang. This work was supported in part by the Research Grants Council of the Hong Kong SAR Government under grant HKUST692/96P.

References:

Jing X.N. *et al.* (1996), *Phys. Rev. B* **53**, 14023.
Pakhomov A.B., Yan X and Zhao B (1995), *Appl. Phys. Lett.* **67**, 3497.
Wong S.K. *et al.* (1998), *preprint*.
Zhao B and Yan X (1997), *J. Appl. Phys.* **81**, 5527, and references therein.

6 Structure of nc-Si:H Films

Gaorong Han, Xiwen Zhang, Guohua Shi, Dake Shen, Weiqiang Han and Piyi Du
State Key Laboratory of Silicon Materials Science, Department of Materials Science and Engineering, Zhejiang University, Hangzhou 310027, P. R. China

Abstract

Hydrogenated nanocrystalline silicon (nc-Si:H) films were prepared by a glow discharge plasma chemical vapor deposition system using heavy hydrogen diluted silane. The structural properties of the deposited films were evaluated by means of Raman spectrometer, high resolution electron microscopy and x-ray diffraction pattern. Effects of the gas flow ratio of SiH_4/H_2 and RF power density on the crystallization and microstructure of nc-Si:H films are described. An interpretation of these results is suggested.

1. INTRODUCTION

In recent years, more and more attention has been paid to the investigation of hydrogenated nanocrystalline silicon (nc-Si:H) films with the development of nano-crystalline materials. Series of studies performed by several research groups (Furukawa and Miyasato, 1988, Takaga et al., 1990, Fortunate et al., 1989, Konuma et al., 1987, He et al., 1993, Martins et al., 1998) have shown that many interesting electrical and optical properties of nc-Si:H films, such as widening of optical bandgap, visible photoluminescence effect and resonant tunneling effect, depend mainly on the volume fraction of silicon crystalline phase and on the crystallite size. However, detailed crystallization kinetics and microstructures of nc-Si:H have not yet been well studied. The aim of the present paper is to show the effect of deposition parameters on the structural properties of nc-Si:H films and to give an interpretation of these results.

2. EXPERIMENTAL

Samples of nc-Si:H were prepared by a conventional capacitively-coupled glow discharge plasma chemical vapor deposition system using SiH_4-H_2 gas mixture. The typical deposition parameters were chamber pressure of 70Pa, substrate temperature of 593K, RF power density of $0.86w/cm^2$, the gas flow ratio of SiH_4/H_2 of 3%, and gas mixture flow rate of 50SCCM. To understand the effect of plasma condition on the formation of silicon nano-scale crystallites, two series of nc-Si:H films were deposited by changing the RF power density from $0.40W/cm^2$ to $1.0W/cm^2$ and the gas flow ratio of SiH_4/H_2 from 2% to 10%. Each sample in thickness of around $1\mu m$ was deposited on the silica glass substrate as well as the polished silicon wafer.

Structural properties of the deposited films were determined by X-ray diffraction (XRD, D/max-RB), Raman spectrometer (Nicovet, Raman 950), and high resolution transmission electron microscope (HREM, JEOL-2010). The volume fraction (ρ) of silicon nanocrystalline phase and mean crystallite size (δ) in each films were estimated from integrated intensity and peak shift of the measured silicon Raman lines.

3. RESULTS AND DISCUSSION

Figure 1 shows the volume fraction (ρ) of silicon crystalline phase and the mean size (δ) of silicon crystallites of the films deposited at different gas flow ratio (X_g) of SiH_4/H_2. As shown in the figure, both of the volume fraction (ρ) and the mean crystallite size (δ) reveals a strong dependence on X_g, reaching to a maximum value at around 3% and then decreasing with the increase of X_g . When X_g increases from 3% to 5%, the volume fraction decreases from 65% to 14% and the mean crystallite size decreases from 9.5 to 4 nm. While $X_g > 6\%$, the deposited films are hydrogenated amorphous silicon (a-Si:H) films.

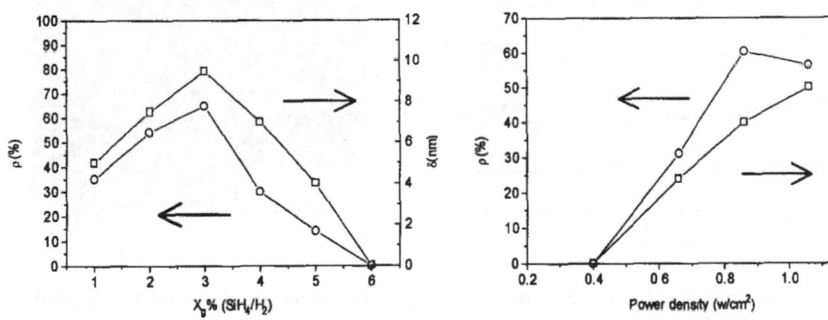

Fig. 1. The volume fraction (ρ) and the mean size (δ) of silicon crystallites in nc-Si:H films as a function of gas flow ratio (X_g)

Fig. 2. The volume fraction (ρ) and the mean size (δ) of silicon crystallites in nc-Si:H films as a function of RF power density

Figure 2 shows the dependence of ρ and δ of the deposited films on RF power density. It is observed that the volume fraction and the mean size of silicon crystallites increase with the increasing of RF power density, which suggests that the structure network of the deposited films changes from amorphous state to crystalline state with the increasing of RF power. When the RF power density is lower than 0.4 W/cm², the deposited film is amorphous.

Furthermore, crystalline structures of the deposited films were determined by the X-ray diffraction. Figure 3 shows the typical XRD spectra of the films deposited on Si (100) substrates at different RF power density, it clearly demonstrates again amorphous state to crystalline state transition of the structure network of the deposited films with the increasing of RF power density. In the XRD spectrum of nc-Si:H film deposited at RF power density of 0.86W/cm², there is an important anomalous phenomenon, that is two extra Bragg reflection peak at 2θ =29.3° and 2θ =32.5° of silicon crystallites except the normal peaks at 2θ =28.5° of Si(111) and 2θ = 47.3° of Si(220). It suggests that the crystallography structure of nano-sized silicon crystallite is different from the common diamond structure of silicon crystal. We have attributed this lattice distortion to the large surface compression pressure resulted from high specific surface tension (Han et al., 1996). Figure 4 illustrates the HREM plane-view images of the same nc-Si:H film sample deposited at RF power density of 0.86 W/ cm² as for XRD measurement. The lattice images reveal that a large quantity of crystalline particles is formed. It can be calculated that sizes of the crystallite range from 3 to 12 nm, and the volume fraction of the crystallite is more than 60%, which is consistent with the results measured by Raman spectra. In addition, we (Han et al., 1996) have observed detailed microstructures of nc-Si:H films by means of HREM and STM. The results have shown that the whirlpool structure, chain structure, pentagon ring structure, quadrilateral and hexagon ring structures exist in interface region of nc-Si:H films.

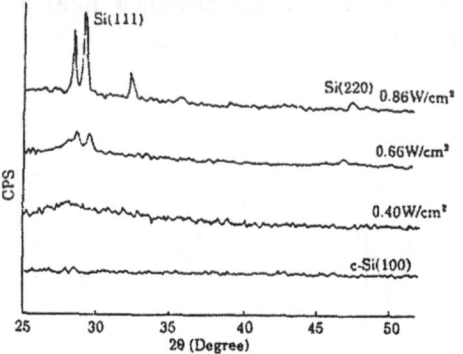

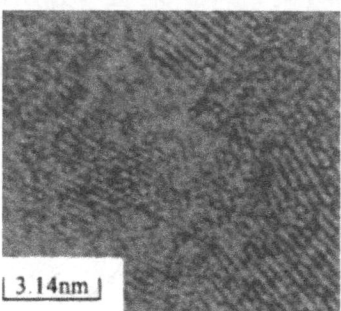

Fig.3. X-ray diffraction spectra of the films deposited at different RF power density

Fig.4. HREM images of nc-Si:H film deposited at RF power density of 0.86 W/ cm²

According to the results presented above, the high quality nc-Si:H film can be prepared at lower gas flow ratio and higher RF power density. The crystallization process of nc-Si:H films could depend on the plasma condition. For the deposition at large gas flow ratio and low RF power density, the SiH₂ neutral free radicals *is* highly concentrated in the plasma and they insert into the Si-H bond on the growing film surface, producing an amorphous silicon (a-Si:H) film (Qiao et al., 1985). When the deposition at small gas flow ratio and high RF power density, the Si and H neutral free radicals become dominated in the plasma, the film growth process is much more Si PVD-like (Tsai et al., 1986), which results in a large amount of crystallites, then, the hydrogenated nanocrystalline silicon (nc-Si:H) film is formed.

4. CONCLUSION

The dependence of the gas flow ratio of SiH_4/H_2 and the power density on the crystallization and the microstructure of nc-Si:H films are obtained. The results show that the high quality nc-Si:H film with the crystallite sizes below 10 nm and crystallite volume fraction above 60% can be prepared at lower gas flow ratio of SiH_4/H_2 and higher RF power density. An interpretation of these results has been given.

ACKNOWLEDGEMENTS

This work is supported by National Natural Science Foundation of China (No. 69890230)

REFERENCE

Fortunate, E., Martins, R., Ferreira, I., Santos, M., Marcario, A., and Guimaraes, L., 1989, J. Non-cryst. Solids, 115, pp.120~122
Furukawa, S. and Miyasato, T., 1988, Phys. Rev. B, 38, pp. 5726~5729
Han, W., Han, G., and Ding, Z., 1996, Chinese Journal of Semiconductor, 17,pp. 406~409
Han, W., Han, G., Ding, Z., Wei, S., Mao, Z., and Zhang, J., 1996, Chinese Science Bulletin, 41, pp. 1139~1142
He, Y., Chu, Y., Lin, H., and Jiang S., 1993, Chinese Phys. Lett., 10, pp. 539~542
Konnma, M., 1987, Philosophical Magazine B, 55, pp. 377~389
Martins, R., Ferreira, I., Fernandes, F., and Fortunato, E., 1998, J. of Non-crystalline Solids, 227-230, pp. 901~905
Qiao, J., Jiang, Z., and Ding, Z., 1985, J. Non-Crystalline Solids, 77&78, pp. 829~832
Takagi, H., Ogawa,H., Yamazaki, Y., Ishizaki, A., and Nakagiri, T., 1990, Appl. Phys. Lett., 56, pp.2379~2380
Tsai, C., Knigts, J., Chang, G., and Wacker, B., 1986, J. Appl. Phys., 59, pp. 2998~3001

7 Nano-oxidation Kinetics of Titanium Nitride Films

S. Gwo[1], T.-T. Lian, Y.-C. Chou and T. T. Chen
*Department of Physics, National Tsing Hua University, Hsinchu 30043,
Taiwan, R.O. C. [¹E-mail: gwo@phys.nthu.edu.tw]*

Abstract

In this work, we examine nano-oxidation kinetics of titanium nitride films with a
conductive-probe atomic force microscope (AFM). The variation of the oxide
height with both sample bias and speed of the AFM tip is found to be significantly
different from the predictions of the classical Cabrera-Mott theory, traditionally
applied to describe very thin oxide film formation on metals.

1 INTRODUCTION

Proximal probe induced anodic oxidation has become a promising new
nanofabrication process that is capable of producing surface oxide patterns on the
nanometer scale (Dagata, 1995). At present, atomic force microscope and scanning
tunnelling microscope (AFM/STM) based anodization of metal films (Matsumoto,
1997) and Si (Teuschler *et al.*, 1995) has been reported. Local field-induced
oxidation (FIO) process is similar to conventional electrochemical anodization
except that an AFM/STM tip is used as the cathode and water from the ambient
humidity is used as the electrolyte. Due to the nature of this process, a
metallurgically stable and chemically resistant film is preferred as the base material.

Titanium nitride (TiN), exhibiting properties of good thermal and electrical
conductivity, extreme hardness, and excellent metallurgical reliability at high
temperatures, is an ideal candidate material for the proximal-probe based
lithographic applications in air. Recently, we have demonstrated the ability to
perform nanolithography of TiN films grown on $SiO_2/Si(100)$ using the local FIO
process near the conducting AFM probe (Gwo, 1999). Here we present our recent
experimental data, indicating that the oxidation kinetics on TiN is significantly
different from the classical Cabrera-Mott (CM) theory (Cabrera and Mott, 1949)
based on a field-enhanced ionic transport mechanism.

2 EXPERIMENTAL

TiN thin films of ≈5 nm thickness were deposited by reactive sputtering technique from a pure Ti target onto Si(100) wafers covered with 100 nm of thermal SiO_2 in an atmosphere of a 1:1 Ar/N_2 gas mixture. The gas flow rate (30/30 sccm) of the sputtering gas was adjusted to give a working pressure of 3.5–4 mTorr. During deposition, the sputtering power was 0.5 kW and the deposition rate was 0.1 nm/sec. The average surface roughness of the deposited films as measured by AFM is around 0.2–0.3 nm.

Local oxidation was performed in air using boron doped p^+-Si cantilevers (10 nm average tip radius) and a commercial AFM/STM microscope (*CP* type, Park Scientific Instruments, CA). The average force constant and resonance frequency of the cantilevers used (Ultralevers ™) are about 0.26 N/m and 40 kHz, respectively. When performing local oxidation, a positive voltage to the sample with respect to the tip was applied at pre-selected lines (so-called "vector" mode) while scanning the sample surface.

3 RESULTS AND DISCUSSION

To understand the FIO kinetics on TiN, we have performed oxide line writing with varied simple biases at a fixed writing speed and with varied writing speeds at a fixed sample bias. Figure 1(a) shows the dependence of the oxide height (H) versus the sample bias voltage (V_s) with a fixed writing speed (0.1 μm/sec). In Fig. 1(b), the averaged cross-sectional oxide height and full-width-half-maximum (FWHM) width (obtained by averaging along individual oxide lines) is plotted versus sample bias. Data in the high electric field regime (Gwo, 1999) show a highly non-linear behaviour. However, in this range of low sample bias (+4–8 V), oxide profile shows a relatively linear relationship with respect to the applied sample bias.

Figure 2(a) shows the dependence of the oxide height (H) versus the writing speed (v) with a fixed sample bias (+8 V). In Fig, 2(b), we plot the H vs. v relation with a wide dynamic range. In the traditional thin film experiments, the oxidation kinetics is studied by plotting the oxide thickness as a function of oxidation time. To get H versus the oxidation time relationship in this case, we take the approximation that the oxidation time t_{ox} equals to the FWHM width of the oxide line w_{ox} divided by the corresponding speed v. The resulting plot is shown in Fig. 2(c).

The most cited kinetic model to describe the FIO process for the case of growing very thin films was first proposed by N. Cabrera and N. F. Mott in 1949. In this model, the strong electric field resulted from the applied bias across the film enhances the injection of metallic cations from the substrate or the injection of oxidant anions (in our case, most likely the OH- ions) from the oxide surface into the oxide film. In this model, the electric field E is assumed to be uniform ($= V_s/H$) throughout the thin film and the growth rate is limited by the decay of the electric field E with increasing height H.

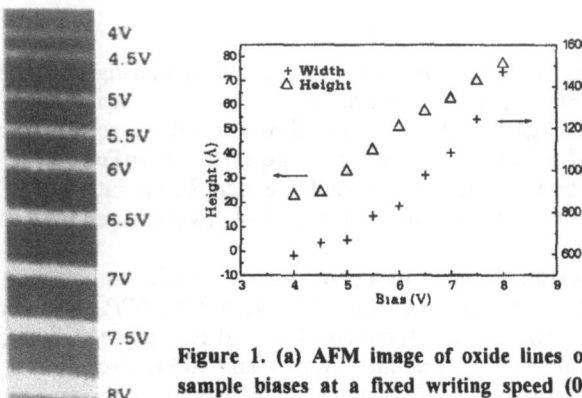

Figure 1. (a) AFM image of oxide lines on TiN obtained with different sample biases at a fixed writing speed (0.1 μm/sec). (b) Cross-sectional height (*H*) and FWHM width (w_{ox}) of oxide lines vs. sample bias (V_s) plot.

In the CM theory, if $H \ll H_l = q(V_0+V_s)a/2k_BT$, with V_0 equal to contact potential, q the electric charge of a mobile ion, k_B the Boltzman constant, T the temperature, and a the mean jump-distance (e.g., distance between two interstitial sites), an inverse logarithmic growth law for metal oxides is predicted, i.e.,

$$\frac{1}{H} = A - B \log t_{ox},$$

where A and $B > 0$ are fit parameters, which contains information about the energy of formation of a mobile ion, the electric field and the energy barrier for diffusion across the oxide. Figure 2(d) displays $1/H$ as a function of $\log(1/t_{ox})$. At room temperature and at the sample bias we used, H_l is in the order of magnitude of several tens of nm. Therefore, $H \ll H_l$ condition holds and this plot can serve as a good way to examine the validity of the CM theory for the TiN case.

Our results suggest that TiN film exhibits a significantly different oxidation kinetics, possibly due to the graded chemical composition when forming oxide layer on binary TiN compound films during the FIO process. It has been shown by us (Gwo, 1999) that the oxidation layer of TiN produced by AFM anodization contains a graded TiN_xO_y transition layer. As a consequence, the electric field is not uniform across the oxide layer. In addition, the ambient FIO process of TiN involves complicated bi-directional transport due to incorporation of O and loss of N across the transition and oxide layers.

4 SUMMARY

We have used the AFM nano-oxidation method to study the growth kinetics of very thin oxide growth on TiN. The typical linear relationship of the oxide height with the applied bias (in the high electric field regime) and the inverse logarithmic of the tip speed [1/log(*v*)] (Stiévenard *et al.*, 1997) is found to be not valid for the TiN case. This work was supported in part by the National Science Council (Contract Nos. NSC 87-2112-M-007-003 and NSC 88-2112-M-007-002), Taiwan, R.O.C.

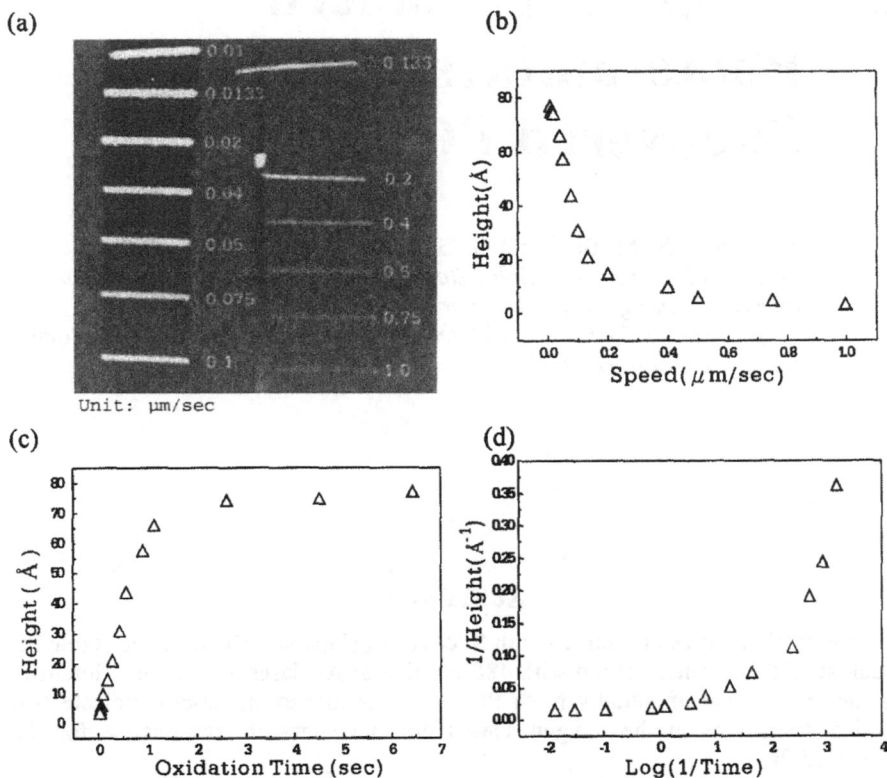

Figure 2. (a) AFM image of oxide lines on TiN obtained with different writing speeds at a fixed sample bias (+8 V). (b) Oxide height as a function of the writing speed. (c) Oxide height as a function of the oxidation time ($t_{ox} = w_{ox}/v$). (d) Inverse oxide height as a function of $\log(1/t_{ox})$.

Cabrera, N. and Mott, N. F., 1949, Theory of the oxidation of metals. *Reports on Progress in Physics*, **12**, pp.163–184.

Dagata, J. A., 1995, Device fabrication by scanned probe oxidation. *Science*, **270**, pp. 1625–1626.

Gwo, S., Yeh, C.-L., Chen, P.-F., Chou, Y.-C., Chen, T.T., Chao, T.-S, Hu, S.-F. and Huang, T.-Y., 1999, Local electric-field-induced oxidation of titanium nitride films. To appear in *Applied Physics Letters*.

Matsumoto, K., 1997, STM/AFM nano-oxidation process to room-temperature-operated single-electron-transistor and other devices. *Proceedings of the IEEE*, **85**, pp. 612–628.

Stiévenard, D., Fontaine, P. A. and Dubois, E., 1997, Nanooxidation using a scanning probe microscope: An analytical model based on field induced oxidation. *Applied Physics Letters*, **70**, pp. 3272–3274.

Teuschler, T., Mahr, K., Miyazaki, S., Hundhausen, M. and Ley, L., 1995, Nanometer-scale field-induced oxidation of Si(111):H by a conducting-probe scanning force microscope: Doping dependence and kinetics. Applied Physics Letters. **67**, 3144–3146.

8 Origin of Enhanced Photoluminescence in Si-covered POPS

X. L. Wu[1], X. M. Bao[1] and G. G. Siu[2]

[1]*National Laboratory of Solid State Microstructures and Department of Physics, Nanjing University, Nanjing 210093, P. R. China and*
[2]*Department of Physics and Materials Science, City University of Hong Kong, Kowloon, Hong Kong*

ABSTRACT

Partially oxidized porous Si with Si covering show an enhanced and stable PL peak at 1.78 eV, when excited with 488 nm line of Ar^+ laser. Its maximal intensity is larger than that of initially fresh PS. Spectroscopic examinations indicate that optical transitions in the oxygen-related defect centers are responsible for the observed PL.

In 1990, Canham (1990) discovered room-temperature strong visible photoluminescence (PL) from porous silicon (PS) fabricated electrochemically. This work was believed to be "opening the door of silicon-optoelectronic integration". However, further investigations indicated that PS has many disadvantages in application. Therefore, improving its stability becomes a subject of numerous investigations. At the end of 1996, Hirschman *et al.* (1996) showed that if PS is oxidized in diluted oxygen at temperature of 700-900 °C, its thermal and chemical stability can be greatly enhanced while retaining desirable light-emission and charge transport properties. However, the emission mechanism of this kind of partially oxidized PS (POPS) material is unclear and the emission efficiency needs to be further improved. In this work, we deposited Si or Ge thin layer on this kind of POPS material. Some new significant results were obtained. We will discuss all the results.

Substrates were <100>-oriented p-type silicon wafers with 1-3 Ωcm resistivity. Electrochemical etching was performed in an electrolyte of $HF:C_2H_5OH=2:1$ with a current density of 15 mA/cm² for 30 min. The thickness and porosity were about 20 μm and 80%, respectively. POPS samples were fabricated by annealing the fresh PS in diluted oxygen (11% in N_2) at 800 °C for 20 min. The oxidized PS was semi-transparent, showing a stable PL peak at 1.72 eV with intensity reduced by a factor of 5. Si or Ge covering was performed using KrF excimer pulsed irradiation. During deposition, the POPS samples were put on a vacuum chamber ($<10^{-5}$ Torr) and kept at about 650 °C. The thicknesses of the deposited layers are in the range of 4-15 nm (about 30-200 pulsed numbers). PL and Raman measurements were run on SPEX 1403 Raman spectrometer, using the 488nm line of Ar^+ laser as excitation beam. All spectra were obtained at room temperature.

Figures 1(a)-1(d) show the PL spectra of a series of Si-covered POPS samples. The spectrum (a) displays a broad PL peak at 1.72 eV. After Si covering, the PL peak shifts to 1.78 eV, exhibiting a blueshift of ~60 meV. Its intensity is evidently enhanced, reaching a maximum in 150Si sample. The maximum is seven times larger than that of the POPS sample and exceeds the PL intensity of the fresh PS. The PL is very stable in air at room temperature. For Ge-covered samples, the maximal PL intensity is only enhanced about 2.7 times and no noticeable blueshift was observed.

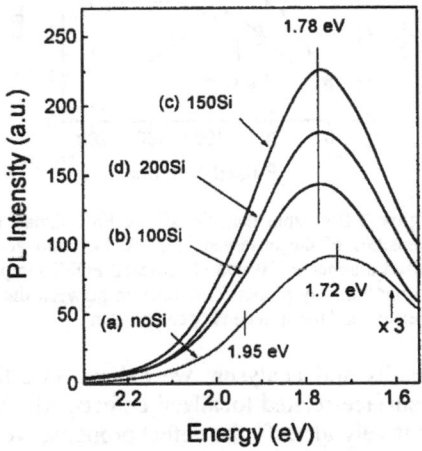

Figure 1 PL spectra of a series of Si-covered POPS samples. (a) POPS, (b) 100Si, (c) 150Si, and (d) 200Si.

The PL blueshift from 1.72 to 1.78 eV cannot arise from the strain resulting from the oxidation of PS surface and subsequent Si covering. Our x-ray diffraction results indicate that the blueshift of the PL peak caused by the strain is at least 0.25 eV (Friedersdorf *et al.*, 1992). Whether does the blueshift arise from the reduction of the Si crystallite sizes? Further Raman analysis indicates that if the 1.72 eV PL was the result of band-to-band recombination in quantum confined Si crystallites, the crystallite size decreases from 20 (noSi) to 13 nm (100Si), the PL peak should blueshift to ~1.734 eV rather than 1.78 eV. Therefore, the quantum confinement model cannot explain the blueshift. In addition, we can also rule out so-called surface state models (Koch *et al.*, 1994; Kanemitsu *et al.*, 1994) having any significant effect on the observed PL.

In fact, Si(Ge) covering on the POPS surface mainly leads to changes of the surface chemistry. Therefore, we draw our attention to the surface chemistry by examining the infrared spectra of our samples. For Ge-covered samples, no noticeable spectral changes were observed. However, for Si-covered samples, spectral changes were very pronounced. A sharp feature is that the 922 cm^{-1} band blueshifts to 977 cm^{-1} and its linewidth narrows. The 920 cm^{-1} mode was assigned to the Si-O-Si symmetric stretching vibration of SiO_2, whereas the 977 cm^{-1} mode to the vibration of stoichiometric SiO (Augustine *et al.*, 1995). Therefore, Si-covering leads to the existence of various SiO bonds and formation of some new compositions between Si and SiO_2.

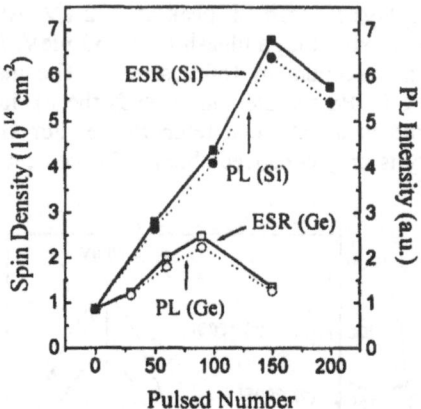

Figure 2 Both spin densities of the ESR signals and intensities of the observed 1.72 eV PL band versus pulsed number in Si- and Ge-covered POPS samples. A quantitatively proportional relation between the PL intensity and the spin density can be seen.

From the about results and analyses, we believe that the enhanced PL is closely associated with surface-related localized defects, which are formed during Si deposition. To quantitatively identify the defect property, we carefully examined the ESR spectra of all Si(Ge)-covered POPS samples. For the PS sample, two weak resonance lines can be observed. The ESR line in the left side was frequently observed in a slightly oxidized PS and attributed to the Si dangling-bond (P_b) signal (Prokes *et al.*, 1994a). The P_b signal disappears in our POPS sample and is slightly enhanced in Si-covered samples. For the right-side resonance line, it weakly appears in our PS and slightly rises in the POPS sample. After Si(Ge) covering, it is further enhanced, reaching an intensity maximum in 150Si(90Ge) samples. If we estimate its spin density by comparing the resonance area with that of a known standard sample, the total spin density versus the pulsed number can be obtained, as shown in Figure 2. Obviously, the spin density tracks very well with the observed 1.72 eV PL intensity. This result implies that some kind of paramagnetic defects is responsible for the 1.72 eV PL. We assigned the defects to the nonbridging oxygen hole centers (NBOHCs) based on the following reasons: 1) The strong ESR signal has a Lande g value of 2.003, consistent with that of the NBOHCs (Prokes *et al.*, 1994a). 2) Its spin density tracks very well with the observed PL intensity, characteristic of the NBOHC-like defects. 3) The NBOHCs defects can reach a local equilibrium by annealing in an inert atmosphere.

resulting in a 60 meV blueshift of the emission energy (Prokes *et al.*, 1994b), in agreement with our experimental observation.

Three different types of NBOHCs have been identified in high purity silica glasses, which vary in PL energy and quantum efficiencies (Prokes *et al.*, 1994b). The first type of NBOHC is Si-O⁻. The second type of NBOHC is Si-O⁻-------H-Si, which is stabilized with a hydrogen bond. The third type of NBOHC is caused by the strain of bonding at an interface between two materials of different bond lengths, density, or structure. The PL intensity caused by the third type of the NBOHCs is proportional to the magnitude of the strain. For Ge(Si)-covered samples, the second type of NBOHCs may be expected to have less influence on the PL intensity. Since Si-O bond is more stable than Ge-O bond, Ge covering should also have small contribution to the first type of NBOHCs. Therefore, the enhancement of the PL intensity in Ge-covered samples mainly comes from the third type of NBOHCs. This is in agreement with our experimental result. For Si-covered samples, the PL intensity caused by the third type of NBOHCs is enhanced about two times, while the first type of NBOHCs leads to the PL intensity enhanced about 5 times. Thus, Si covering mainly gives rise to the introduction of the first type of NBOHCs. Since the Ge(Si)-covered samples were fabricated at temperature of 650 °C, the stability of the observed PL is understandable. Our experimental results provide a way for improving stable PL, which will be more useful for device applications.

This work was supported by Natural Science Foundations of China (No.59832100) and Jiangsu Province as well as climbing program.

References

Augustine, B. H., Irene, E. A., He, Y. J., Price, K. J.,.McNeil, L. E., Chrisensen, K. N. and Maher, D. M., 1995, Visible light emission from thin films containing Si, O, N, and H, *Journal of Applied Physics*, **78**, 4020-4030.

Canham, L. T., 1990, Silicon quantum wire array fabrication by electrochemical and chemical dissolution of wafers, *Applied Physics Letters* **57**, pp 1043-1045

Friedersdorf, L. E., Searson, P. C., Prokes, S. M., Glembocki, O. J., and Macaulay, J. M., 1992, Influence of stress on the photoluminescence of porous silicon structures, *Applied Physics Letters* **60**, pp 2285-2287

Hirschman, K. D., Tsybeskov, L. S., Duttagupta, P. and Fauchet, P. M., 1996, Silicon-based visible light-emitting devices integrated into microelectronic circuits, *Nature* **384**, pp 338-341

Koch, K. and Petrova-koch, V., 1994, The surface state mechanism for light emission from porous silicon, in. *Porous Silicon*, edited by Z. C. Feng and R. Tsu (World Scientific, Singapore), pp. 133-148

Kanemitsu, Y., Matsumoto, T., Futagi, T. and Mimura, H., 1994, Microstructure: Optical properties and application to light emission diodes, in *Porous Silicon*, edited by Z. C. Feng and R. Tsu (World Scientific, Singapore), pp. 363-392.

Prokes, S. M., Carlos, W. E. and Glembocki, O. J., 1994a, Defect-based model for room-temperature visible photoluminescence in porous silicon, *Physical Review B* **50**, pp 17093-17096

Prokes, S. M. and Glembocki, O. J., 1994b, Role of interfacial oxide-related defects in the red-light emission in porous silicon, *Physical Review B* **49**, pp 2238-2242.

9 Spin Split Cyclotron Resonance in GaAs Quantum Wells

X. G. Wu
NLSM, Institute of Semiconductors, Chinese Academy of Sciences

Abstract

The cyclotron resonance (CR) of electrons in GaAs/AlGaAs quantum wells is investigated theoretically to explain a recent CR experiment, where two CR peaks were observed at high magnetic fields when both spin-up and spin-down states of the lowest Landau level are occupied. Our theoretical model takes into account the conduction band non-parabolicity, the electron bulk longitude-optic-phonon coupling, and the self-consistent subband structure. A good agreement is found.

In a semiconductor with a non-zero effective g-factor and a non-parabolic conduction band, e.g., GaAs, one expects to see spin splitting of the cyclotron resonance (CR), as electrons with different spin orientations have different cyclotron frequencies. For quasi two-dimensional electron systems (2DES) formed in GaAs/AlGaAs heterostructures and quantum wells, this spin splitting has attracted considerable interests recently.

In some experiments (Michels et al., 1993), only a single CR peak can be resolved, although samples have very high electron mobility. In some other experiments (Summers et al., 1993, Hu et al., 1995), a spin split CR was observed, but in samples with an order of magnitude lower electron densities. It was found that, as the electron density increases, two CR peaks merge into a single one, and this has been attributed to the electron-electron interaction (Cooper and Chalker, 1993).

In a recent experiment (Wang et al., 1998), it is observed in high electron density samples that, at low magnetic fields, below the reststrahlen region, only a single CR peak can be identified, consistent with previous experiments (Michels et al., 1993). However, at high magnetic fields, above the electron longitude-optic (LO) phonon resonance, two well resolved CR peaks can be seen, when both spin-up and spin-down states of the lowest Landau level are partially occupied.

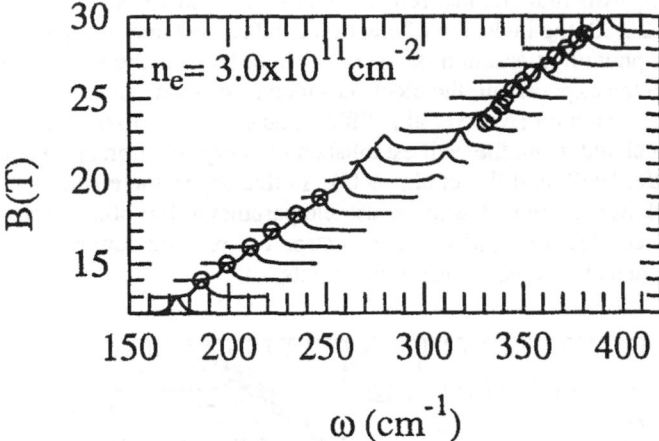

Fig.1 Calculated CR spectra for various magnetic fields for sample A. The experimental CR frequency vs the magnetic field is shown as open circles (Wang et al., 1998).

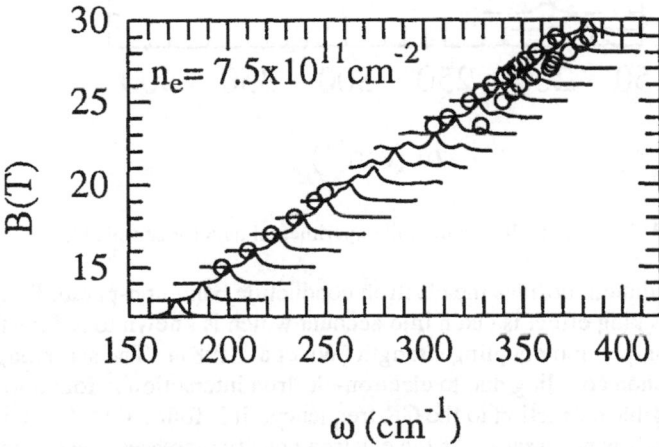

Fig.2 Calculated CR spectra and experimental data for sample B.

In this paper, we wish to report a theoretical study of the electron CR in GaAs quantum wells. A model, which takes into account the conduction band non-parabolicity and the electron bulk LO-phonon interaction, is proposed to explain the experiment findings. This model demonstrates that, the correction to the CR frequency due to electron-phonon interaction strongly depends on the occupation of spin polarized electrons. It generally leads to two CR peaks and the size of splitting varies with the electron occupation. A comparison is made between the theory and experiment, and a good agreement is found.

Previous experimental and theoretical studies have shown that, in analyzing the CR of quasi 2DES, the inclusion of electron conduction band non-parabolicity and the electron-phonon interaction is essential. In order to make a quantitative comparison with the experiment, the electron subband structure must be calculated self-consistently (Ando et al., 1982). The correction to the CR frequency is calculated using the well established memory function approach (Gotze and Wolfe, 1972, and Wu et al., 1987). In this paper, the electron conduction band mass is treated as an adjustable parameter, but other parameters, e.g., the valence-conduction band gap, the electron LO-phonon coupling constant, and the electron density are taken as known inputs.

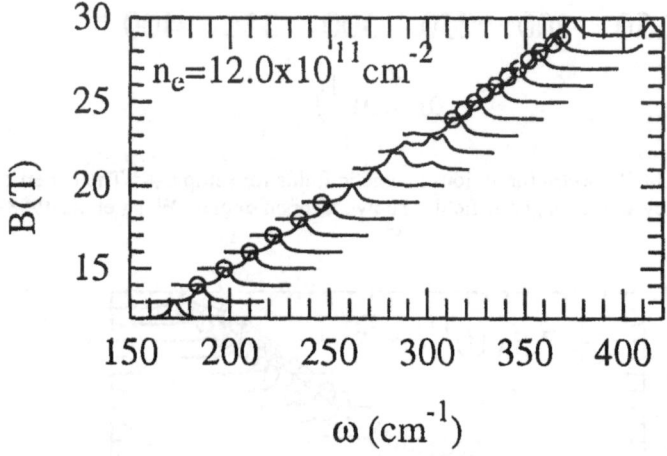

Fig.3 Calculated CR spectra and experimental data for sample C.

In the calculation, besides the electron conduction band non-parabolicity, the Fermi-Dirac blocking effect is taken into account which is known to reduce the effective electron-phonon coupling strength (Wu et al., 1987). The screening of the electron-phonon coupling due to electron-electron interaction is found to produce a negligible correction to the CR frequency. It is found that the inclusion of higher subbands is necessary. As the electron density increases, the subband energy gap decreases (Ando et al., 1982). At high magnetic fields, the CR frequency becomes comparable to the subband gap plus LO-phonon frequency, and this leads to an additional resonant contribution to the memory function, thus a large correction to the CR frequency.

In Figs.1-3, the experimental CR frequency vs the magnetic field is shown as open circles for samples A, B, and C, respectively. In each figure, the calculated CR absorption vs frequency are plotted for different magnetic field ranging from 13T (bottom) to 30T (top) in steps of 1T. Each spectrum is shifted vertically for clarity. The sample electron density is shown in each figure.

In Fig.1, one sees that, a single CR peak is observed. In Fig.2, one can see a single CR peak at low magnetic field region, but a second CR peak appears at high magnetic fields. In Fig.3, again only a single CR can be seen. It is found that the correction to the CR frequency due to electron LO-phonon interaction depends on the magnetic field strength. The higher the magnetic field is, the larger the correction will be. At low field, the splitting is smeared out by the broadening due to, e.g., electron-impurity scattering. The splitting due to conduction band non-parabolicity is found to be rather small, thus the influence of electron-phonon interaction is important. The electron conduction band mass 0.0665 is used throughout, and this is a well-accepted value in the literature. It is clear that the agreement between our theory and the experiment is reasonably good.

X.G. Wu was partly supported by the NSF of China.

References

Ando T., et al., 1982, Rev. Mod. Phys. **54**, p. 437.
Cooper N.R. and Chalker J.T., 1994, Phys. Rev. Lett. **72**, p. 2057.
Gotze W. and Wolfe P., 1972, Phys. Rev. B **6**, p. 1226.
Hu C.M., et al., 1995, Phys. Rev. B **52**, p. 12090.
Michels J.G., et al., 1993, Surf. Sci., **305**, p. 33.
Summers G.M., et al., 1993, Phys. Rev. Lett. **70**, p. 2150.
Wang Y.J., et al., 1998, Proceedings of High Magnetic Fields in the Physics of Semiconductors, Wurzburg, Germany.
Wu X.G., et al, 1987, Phys. Rev. B **36**, p. 9760.

10 The Effect of Intermixing on Tensile-strained Barrier GaAs/GaAsP Quantum Well Structure

Michael C. Y. Chan and E. Herbert Li
Department of Electrical and Electronic Engineering, University of Hong Kong, Pokfulam Road, Hong Kong

ABSTRACT

The tensile-strained barrier GaAs/GaAs$_{1-x}$P$_x$ quantum well (QW) structures fabricated on GaAs substrate have a remarkable potential for novel properties of laser structures. In the case of tensile-strained barrier GaAs/GaAsP QW, a small amount of light-hole (LH) and heavy-hole (HH) splitting is attainable within a large range of well width and P compositions. In order to shift the heavy hole and light hole energy levels, the concept of band-gap engineering is a useful tool to accept the particular devices operation. Basically, intermixing process is one of easy ways to achieve the modification of band structure. The effect of band structure on the intermixed GaAs QW with tensile-strained GaAsP barriers grown on GaAs substrates are reported.

1. INTRODUCTION

Modification of band structures to improve the performance and characteristics of optical devices has been one of the major research areas for band structure engineering. For QW optical devices, combined both the factors of quantum size and strain induced in layered structure allowed a higher flexibility for band structure engineering. This can improve the certain characteristics of laser diode, including higher differential gain, lower threshold current, higher output power, enhanced linewidth and modulation bandwidth (Adams, 1986). At 1992 the tensile-strained QW diode laser has reported with lower current densities, and high temperature operation, and polarization insensitive devices utilizing these structures (Mikami *et al*, 1992).

Heterostructure of GaAs well, GaAsP barrier growth on GaAs substrate provide a tensile-strained barrier structure without involving complicated quaternary alloys (Agahi *et al*, 1994). The P concentration in barrier and the well width is the two elements that can be vary to tune for a particular band structure. One of the structures that researches have great interest on is the merging of the light-hole and the heavy-hole band. The merge of the two bands would enhance electroabsorption, and polarization independence, polarization independence offer advantage to various device applications.

In this paper, we present a detailed theoretical analysis of the interdiffused GaAsP/GaAs tensile-strain barrier QWs. This includes the confinement profiles and LH-HH splitting. In section 2, we will present the diffusion models. In section 3, the as-grown structure is selected to demonstrate how the LH-HH splitting are affected by the technique of intermixing. Finally the conclusions are drawn.

2. THEORETICAL MODEL

The post-growth process of thermally induced composition intermixing of QWs has becoming a popular technique for material tuning and which, is proposed here (Li, 1998). We demonstrated the interdiffusion technique to tune the band structure. The interdiffusion technique is a thermal process consists of consituant atom moving across the hetero-interface. Interdiffusion technique shows its ability to merge both the light-hole and heavy-hole band. One of the major advantages using interdiffusion technique for band structure tuning is because its accuracy is much higher than for growing an precise as-grown heterostructure.

In our diffusion model, it is assumed that Fick's second law is obeyed in the QW layers and all atoms movement through the interface between the adjacent hetro-layers have the same diffusion coefficient. Interdiffusion across the heterointerface alters the composition profile across the QW structure. In GaAsP/GaAs QW structures only the interdiffusion of group-V atoms occurs, i.e. As and P atoms, since there is no Ga concentration gradient across the interface. The composition profile after interdiffusion is characterized by a diffusion length L_d, which is defined as $L_d = \sqrt{(Dt)}$, where D is the diffusion coefficient and t is the annealing time of thermal processing.

Lattice mismatch between thin well and thick barrier QW materials can be taken up by strain and results in a pseudomorphic QW such that an uniform lattice constant can be found through out the whole structure. This tetragonal deformation results in strain and which is, perpendicular to the hetero-interface. The in-plane strain across the well will vary according to the composition of alloy concentration after interdiffusion.

3. RESULTS AND DISCUSSION

The as-grown structure analyzed here is an $GaAs_{1-x}P_x/GaAs$ single QW, with well width L_z=60Å. The structure is assumed to be fabricated on GaAs substrate with GaAs in the well and GaAsP in the barrier forming a tensile-strain barrier QW. The results for various L_d and P concentrations are obtained. In our calculation, the as-grown P concentration in the QW (x_o) is set to be 0.05, 0.2 and 0.4 so as to create different degrees of tensile strains. All the parameters were determined by Vegard's Law between the binary parameters at room temperature. The conduction band offset of GaAsP/GaAs heterostructures is set to be 0.57.

For x_o = 0.2, there is an 0.72 % tensile strain produced in the barrier. As the interdiffusion proceeds, compositional disordering changes the tensile strain accordingly. The tensile strain will also induce in the well. At L_d=20Å, the tensile strain will be 0.2% in the well centre.

The valence band confinement potentials of interdiffused QWs with phosphorous x_o equal to 0.2 are plotted in Fig.1. Owing to the lack of space, results for holes, x_o=0.05 and 0.4 are not shown as the main features are very similar to those in Fig.1. The P concentration in the well increases with L_d as As atoms diffuse into the barriers from the well. When L_d is small, the P concentrations near to the interface drop quickly while the concentration at the centre remains unchanged, which leads to a smaller difference of the confinement potential between HH and LH subbands at well centre. When the L_d reaches 20Å, the confinement potential differences will further increase. The subband-edge energy levels of HHs and LHs of the as-grown square QW are shifted down in the interdiffused QW structures according to the extent of interdiffusion, the zero-reference is taken to be at the top of confinement potential of the as-grown QW. The first (ground) subband-edge energy for HH and LH are shifted far away from the conduction band. The energy are shifted from 12.22 to 40.61meV for HH and from 29.02 to 34.40meV for LH with L_d=0-20Å. This feature is another degree of freedom to merge the HH and LH subbands together to achieve the polarization independent devices.

For the particular combination of alloy composition and QW widths, the merging of HH and LH subband can be achieved for the GaAsP/GaAs QW. When tensile strain is in the barrier layers with well lattice-matched to the substrate, the limitation of the LH-HH splitting is determined by the well width and the amount of tensile strain which can modifies the confinement potential barrier height in the valence band. For the interdiffused QW, the tensile strain induced in the well layer which can also modifies the confinement potential. The changes of the LH-HH splitting are also dependent on the interdiffusion. Fig. 2 shows the splitting of LH-HH as the function of L_d for the three P compositions. The L_d over 10Å is required to merge the LH-HH splitting.

4. CONCLUSION

In summary, we have presented the interdiffusion of tensile-strain barrier GaAs$_{1-x}$P$_x$/GaAs single QW with 60Å well width. By a suitable choice of the diffusion parameters, it is possible to cause the coincidence of energy levels of HH and LH resulting in polarization-independent operation devices. This creation of tensile strain in the well through interdiffusion results in the merging of HH and LH subbands.

ACKNOWLEDGEMENT

This paper is supported by the RGC earmarked grant of Hong Kong and the University of Hong Kong CRCG research grant.

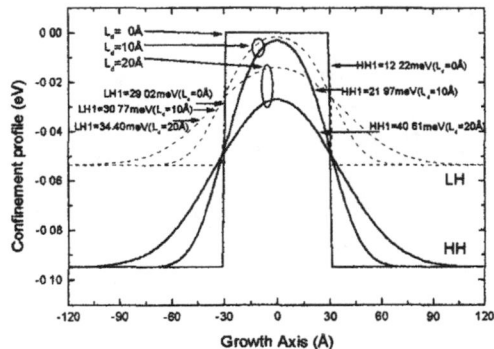

Fig.1 The confinement potential profiles of the heavy hole (HH) and light hole (LH) subbands with different diffusion lengths

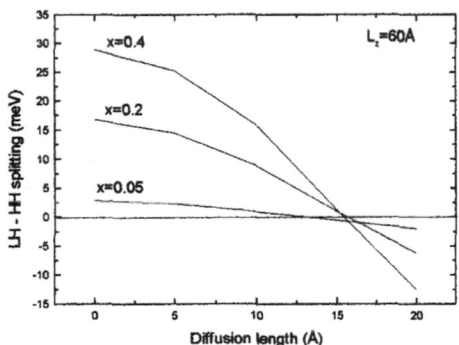

Fig.2 The LH-HH splitting as a function of diffusion lengths for three P compositions.

REFERENCE

Adams, R., 1986, Band structure engineering for low-threshold high effeciency semiconductor lasers. Electron. Lett., **22**, pp. 249-250.

Agahi, F., Lau, K. M., Koteles, E. S., Baliga, A., Anderson, N. G., 1994, GaAs$_{1-x}$P$_x$/GaAs Quantum-Well Structures with Tensile-Strained Barriers, IEEE J. Quantum Electron., **30**, pp. 459-465.

Li, E. H., 1998, Quantum Well Intermixing for Photonics, (Milestone Series, vol. 145).

Mikami, Noguchi, Y., Magari, K., and Suzuki, Y., 1992, Polarization insensitive superluminescent diode at 1.5 mm with a tensile-strained barrier MQW, IEEE Photon. Technol. Lett., **4**, pp. 703-705.

Part 2

CARBON
NANOSTRUCTURES

11 Carbon Nanotubes: Interactions, Manipulation, Electrical Properties and Devices

Phaedon Avouris and Richard Martel
IBM Research Division, T. J. Watson Research Center, Yorktown Heights, NY 10598, USA

ABSTRACT

Carbon nanotubes are novel materials with unique electrical and mechanical properties. Here we present our results on their structure and interactions in the adsorbed state, on their self-organization into rings, on ways to manipulate individual nanotubes, on their electrical properties and how these properties are affected by structural distortions, and, finally, on the fabrication and characteristics of nanotube-based electron devices. Specifically, atomic force microscopy (AFM) and molecular mechanics simulations are used to investigate the effects of van der Waals interactions on the atomic structure of adsorbed nanotubes. Both radial and axial structural deformations are identified and the value of the interaction energy itself is obtained from the observed deformations. We demonstrate that the strong substrate-nanotube interaction allows the AFM manipulation of both position and shape of individual nanotubes at surfaces. We then concentrate on the electrical transport properties of nanotubes. We briefly review the literature and discuss recent theoretical results concerning the effect of structural distortions of nanotubes on their electrical properties. Preliminary results regarding the effects of a magnetic field on the transport of nanotube are also reported. Finally, we demonstrate the operations of a field-effect transistor based on a single semiconducting nanotube, and a single-electron transistor using a nanotube bundle as Coulomb island.

1. INTRODUCTION

Carbon nanotubes (NTs) are an interesting class of nanostructures which can be thought of as arising from the folding of a layer of graphite (a graphene sheet) to form a hollow cylinder composed of carbon hexagons. A number of techniques including discharges between carbon electrodes, laser vaporisation of carbon, and thermal decomposition of hydrocarbons have been used to prepare these materials

(Dresselhaus, 1996; Thess, 1996; Journet, 1998). Depending on the width of the graphene sheet and the way it is folded a variety of different nanotube structures can be formed. As shown in Figure 1.1, a complete description of the nanotube structure is provided by defining a chiral vector $C = n\mathbf{a} + m\mathbf{b} \equiv (n, m)$, where $\mathbf{a}$ and $\mathbf{b}$ are the unit vectors of the two-dimensional graphene sheet and n, m are integers, and a chiral angle θ, which is the angle of the chiral vector with respect to the zig-zag direction of the graphene sheet (Dresselhaus, 1996; Saito, 1998). When the graphene sheet is rolled up to form a particular nanotube, the two ends of the chiral vector are joined, and the chiral vector forms the circumference of the nanotube. In a 2-D graphene sheet the π and π^* states become degenerate at the K-points of the hexagonal Brillouin zone, resulting in a zero-gap semiconductor. In a nanotube, quantization of the wavefunction along its circumference restricts the allowed wave vectors to certain directions of the graphite Brillouin zone so that $C \cdot k_x = 2\pi j$, where j is an integer. If at least one of these wave-vectors passes through a K point, the tube is metallic; otherwise it is a semiconductor with a finite band-gap. Thus, $(n,0)$ zig-zag NTs are expected to be metallic if $n/3$ is an integer; otherwise they are semiconductors (Dresselhaus, 1996; Saito, 1998).

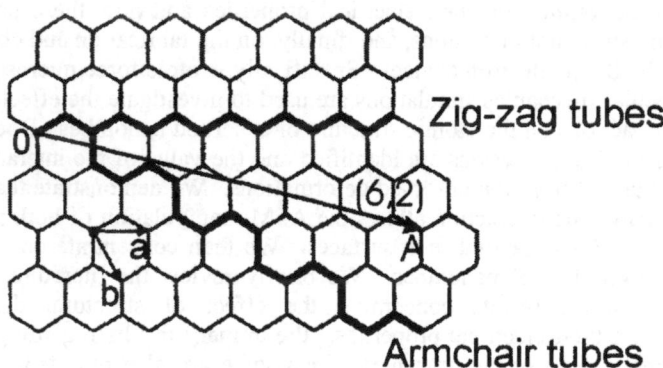

Zig-zag tubes

(6,2)

Armchair tubes

Figure 1.1 A graphene sheet with unit vectors **a** and **b**. A nanotube can be formed by folding the sheet so that the end points 0 and A of the vector $C = n\mathbf{a} + m\mathbf{b} \equiv (n, m)$ coincide. When n=m, the ends of the tube have a meander-like structure and the tube is called am armchair tube. When n=0, the ends of the nanotube have a zig-zag structure. Nanotubes with arbitrary values of n and m have a chiral structure. Adapted from Kuzmany, Phys. Bl. **54**, 331 (1998).

As the chiral vector rotates away from the $(n,0)$ direction, the resulting (n,m) tubes are chiral and are expected to be metallic if $(2n+m)/3$ is an integer. If this is not the case then they are semiconductors with a gap $\propto 1/R$, where R is the NT radius (Mintmire, 1992). Finally, when C is rotated $30°$ away from the $(n,0)$ direction, $n=m$ and the resulting armchair NTs are metallic. Recent scanning tunneling microscope studies have verified the basic conclusions of the simple theory (Wildöer, 1998; Odom, 1998). For NTs of finite length L one also has to consider the quantization of the wave-vector k_y along the tube direction, i.e. $k_y L = q\pi$, where q is an integer. A simple estimate of this effect can be obtained by approximating the dispersion relation at the Fermi energy as $E = E_F - hv_F (k-k_F)$, which leads to a level spacing given by $\Delta E = hv_F / 2L$ (Tans, 1997). Using $v_F \approx 8.1\times10^5$ m/s, the level spacing induced in a $0.5\mu m$ long NT is estimated to be ~3.5 *meV*. Finally, we note that carbon nanotubes can be made of a single layer of carbon atoms (single-walled tubes, SWNTs) or many layers (multi-walled tubes, MWNTs).

The interesting electrical properties described above, coupled with their high mechanical strength (Treacy, 1996; Wong, 1997), give a unique character to these materials. Many applications have been proposed (Dresselhaus, 1996). In particular, nanotubes appear very promising as one-dimensional nanowires and as switching elements in novel nanoelectronic devices.

Here we discuss a number of issues involving the structure, interactions, manipulation and electronics applications of nanotubes. Specifically, we present results on the interaction of NTs with inert substrates. We show that the van der Waals interactions with these substrates can be quite strong, inducing both axial and radial distortions of the atomic structure of nanotubes. The van der Waals binding energy itself can be estimated by measuring the extent of these distortions using atomic force microscopy. We then discuss how one can use the van der Waals forces acting between NTs to generate a new state of NT self-organization namely, nanotube rings. We show that the strong nanotube-substrate interaction makes possible the manipulation of not only the position but also the shape of individual nanotubes at room temperature using the tip of an atomic force microscope (AFM). In this way, for example, NTs can be positioned on electrical contacts so that the electrical characteristics of a single NT can be measured. We will then concentrate on the electrical properties of NTs. Some concepts relevant to transport in low dimensional systems and recent experimental results first are reviewed. Recent theoretical work on the effect of bending of NTs on their electrical properties will be discussed. Preliminary results on the transport properties of nanotube rings and the effects of an external magnetic field on these properties will also be presented. An important finding is that the carriers in a semiconductor NT can be depleted with the help of an external electric field. We will discuss in some detail the fabrication and operating characteristics of field-effect and single-electron transistors based on single NTs. Finally, we demonstrate that the metallic character of nanotubes allows their use as electrodes for the local electrochemical modification of surfaces.

2. NANOTUBE-SUBSTRATE INTERACTIONS: BINDING AND DISTORTIONS

As mentioned in the introduction, the electronic properties of carbon nanotubes can be derived from those of graphite on the basis of symmetry related arguments. Unlike free nanotubes, however, nanotubes supported on a solid substrate may be distorted both axially and radially. Figure 2.1 shows two non-contact AFM images of overlapping multi-wall nanotubes on an H-passivated silicon surface. In this case, the nanotubes are expected to interact with the inert substrate by van der Waals forces. The images clearly show that the upper tubes bend around the lower ones. As we have discussed in detail elsewhere (Hertel, 1998a; 1998b), these distortions arise from the tendency of the upper NTs to increase their area of contact with the substrate so as to increase their adhesion energy. Counteracting this tendency is the increase in strain energy that results from the increased curvature of the upper tubes. The total energy of the system can be expressed as an integral of the strain energy $U(c)$ and the adhesion energy $V(z)$ over the entire tube profile:

$$E = \int [U(c) + V(z(x))]dx \cdot$$

Here, c is the local tube curvature and $V(z(x))$ the nanotube-substrate interaction potential at a distance z above the surface. Using the measured Young's modulus for MWNTs (Treacy, 1996; Wong, 1997) and by fitting to the experimentally observed nanotube profile, one can estimate the binding energy from the observed distortion. In this way we obtain, for example, a binding energy of ~0.8 eV/Å for MWNTs with a diameter of about 100 Å. Thus, van der Waals binding energies, which for individual atoms or molecules are weak (typically 0.1 eV), can be quite strong for mesoscopic systems such as the NTs. High binding energies imply that strong forces are exerted by nanotubes on underlying surface features such as steps, defects, or other nanotubes. For example, the force leading to the compression of the lower tubes in Figure 2.1a is estimated to be as high as 30 nN.

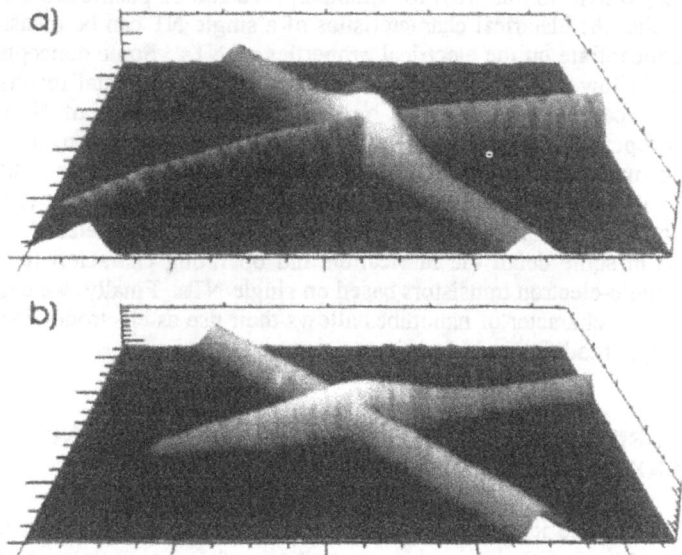

Figure 2.1 Atomic force microscope (AFM) non-contact mode images of two overlapping multi-wall nanotubes. The upper tubes are seen to wrap around the lower ones which are slightly compressed. The size of image (**a**) is 330x330 nm and that of (**b**) is 500x500 nm.

In addition to the experimental study of the distortions induced by van der Waals forces, we have performed molecular mechanics simulations of the same phenomena (Hertel, 1998b). Molecular mechanics represents an alternative to the Born-Oppenheimer approximation-based electronic structure calculations. In this case, nuclear motion is studied assuming a fixed electron distribution associated with each atom. The molecular system is described in terms of a collection of spheres representing the atoms, which are connected with springs to their neighbors. The motion of the atoms is described classically using appropriate potential energy functions. The advantage of the approach is that very large systems (many thousands of atoms) can be simulated. In our simulations, the MM3 alkene force-field was used to model the intra-tube atomic interactions, while the van der Waals interaction parameters were obtained by summing atomic van der Waals interactions for sp^2 hybridized carbon atoms interacting with a graphite slab.

Figures 2.2(a) and (b) show the distortions that arise when two single-walled (10,10) NTs cross each other. In addition to their axial distortion, the two nanotubes are seen to have a distorted, non-circular cross-section in the overlap region. Results on the radial distortions of single-walled nanotubes due to van der Waals interactions with a graphite surface are shown in Figure 2.2(c). It is found that adhesion forces tend to flatten the bottom of the tubes so as to increase the area of contact. At the same time, there is an increase in the curvature of the tube and therefore a rise in strain energy. The resulting overall shape is dictated by the optimization of these two opposing trends. Small diameter tubes that already have a small radius of curvature R_c resist further distortion ($E_{STRAIN} \propto R_c^{-2}$), while large tubes, such as the (40,40) tube, flatten out and increase considerably their binding energy (by 115% in the case of the (40,40) tube). In the case of MWNTs, we find that as the number of carbon shells increases, the overall gain in adhesion energy due to distortion decreases as a result of the rapidly increasing strain energy (Hertel, 1998b).

The AFM results and the molecular mechanics calculations indicate that carbon nanotubes in general tend to adjust their structure to follow the surface morphology of the substrate. One can define a critical radius of surface curvature R_{CRT} above which the nanotube can follow the surface structure or roughness. Given that the strain energy varies more strongly with tube diameter ($\propto d^4$) than the adhesion energy ($\propto d$), the critical radius is a function of the NT diameter. For example, R_{CRT} is about $(12d)^{-1}$ for an NT with a d=1.3 nm, while it is ~$(50d)^{-1}$ for an NT with d=10 nm.

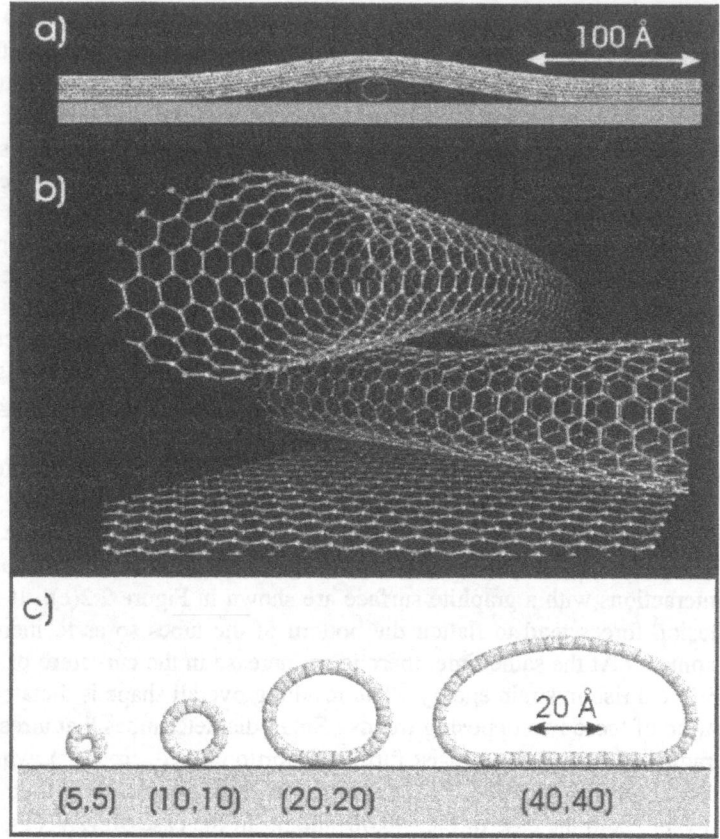

Figure 2.2 Molecular mechanics calculations on the axial and radial deformation of single-wall carbon nanotubes. (a) Axial deformation resulting from the crossing of two (10,10) nanotubes. (b) Perspective close up of the same crossing showing that both tubes are deformed near the contact region. The forces acting on the lower tube is about 5 nN. (c) Computed radial deformations of single-wall nanotubes adsorbed on graphite.

3. SELF-ORGANIZATION OF CARBON NANOTUBES: NANOTUBE RINGS

The van der Waals forces play an important role not only in the interaction of the nanotubes with the substrate but also in their mutual interaction. The different shells of a MWNT interact by van der Waals forces; single-walled tubes form ropes for the same reason. Recently, we have found a way to convert short ropes of SWNT into nanotube rings (Martel, 1998a). This method involves the irradiation of raw SWNTs in a sulfuric acid-hydrogen peroxide solution by ultrasound. This treatment both etches the NTs, shortening their length to about 3-4 μm, and induces ring formation. Figure 3.1 (left) shows a number of such rings while Figure 3.1 (right) shows a two ring catenane structure. Most of the rings appear to be single circular torii as was suggested previously (Liu, 1997). However, TEM images, ring unwinding through AFM manipulation, and the method of ring preparation itself suggest that the rings are not torri structures but coils of many SWNTs.

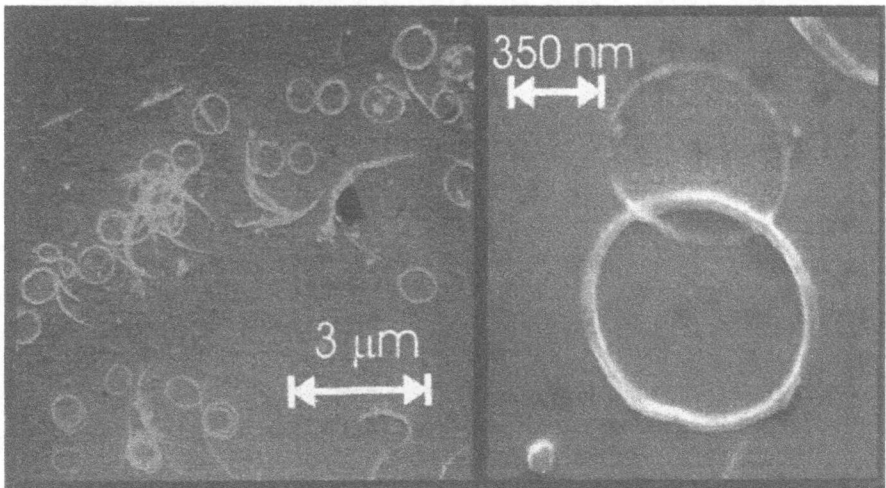

Figure 3.1 Scanning electron microscope images of nanotube rings produced from ropes of single-wall nanotubes. Right: images of two inter-connected nanotube rings.

Coils of biomolecules and polymers are well known structures and involve a number of interactions that include hydrogen bonds and ionic interactions (Bryson, 1995; Creighton, 1992). Nanotubes rings are, however, rather surprising objects given the high flexural rigidity of NTs and the fact that NT coils can only be stabilized by van der Waals forces. We investigated the thermodynamic stability of NT coils using continuum mechanics, and we found that in fact it is easy to compensate for the strain energy induced by the coiling process through tube-tube adhesion in the coil. In Figure 3.2 we show how a given length of nanotube l should be divided between the perimeter of the coil $2\pi R$ that defines the strain energy and the interaction length $l_i = l - 2\pi R$ that contributes to the adhesion (see inset picture) so that a stable structure is formed. This stability plot is computed for three SWNTs with radii 0.7 nm (solid line), 1.5 nm (dotted) and 4.0 nm (dashed). The values of the cohesive energy per unit length were obtained from the calculations of Tersoff and Ruoff (Tersoff, 1994). From Figure 3.2 it is clear that the critical radius R_c for forming rings is small, especially for small radius NTs such as the (10,10) tube (r = 0.7 nm). In general, we observe that much lower radius rings are allowed than are observed experimentally (see histogram of Fig. 3.3). This remains true even if we take into account the hydrophobic interactions between the nanotube and the aqueous environment.

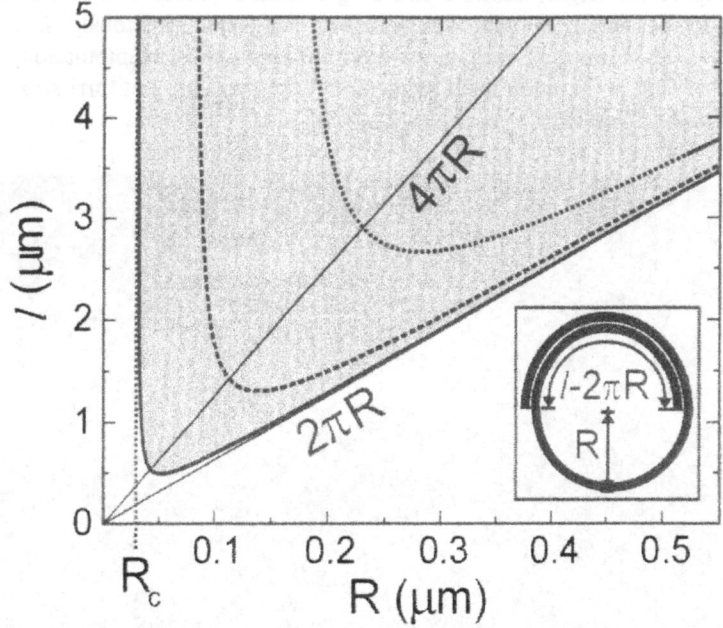

Figure 3.2 Thermodynamic stability limits for rings formed by coiling single wall nanotubes with radii of 0.7 nm (━━), 1.5 nm (••••), and 4.0 nm (••••) calculated using a continuum elastic model.

It is clear that the coiling process is not controlled by thermodynamics but by kinetics. The reason is easy to understand; to form a coil the two ends of the tube have to come first very close to each other before any stabilization (adhesion) begins to take place. This bending involves a large amount of strain energy (E_{STRAIN} $\propto R^2$), and the activation energy for coiling will be of the order of this strain energy (i.e. several eV). Similar arguments hold if, instead of a single SWNT, we use a SWNT rope. Given that NT rings form with a high yield (~50%), how is this high activation energy supplied? In our experiments, the coiling process is driven by exposure to ultrasound. Ultrasonic irradiation can provide thermal activation (Suslick, 1988), however, it is unrealistic to assume that the huge energy needed is supplied in the form of heat energy. It is far more likely that mechanical processes associated with cavitation, i.e. the formation and collapse of small bubbles in the aqueous solvent medium that are generated by the ultrasonic waves, are responsible for tube bending. (Suslick, 1988). The nanotubes may act as nucleation centers for bubble formation so that a hydrophobic nanotube trapped at the bubble-liquid interface is mechanically bent as the bubble collapses. Once formed, a nanotube "proto-ring" can grow thicker by the attachment of other segments of SWNTs or ropes.

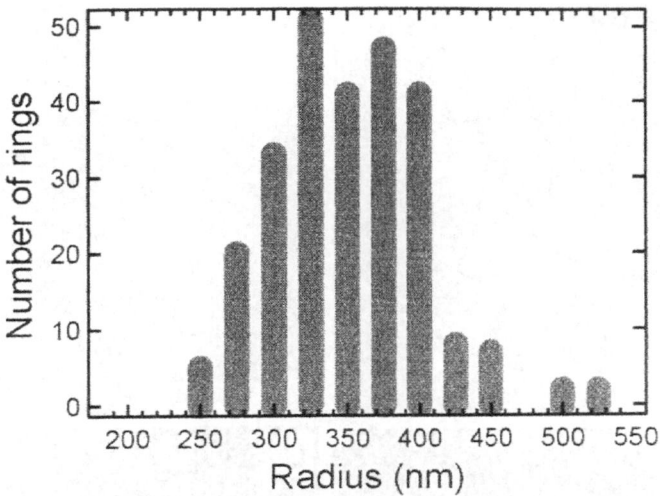

Figure 3.3 Histogram showing the distribution of the radii of nanotube rings. The histogram was obtained by analysing a large number of SEM images of rings.

4. MANIPULATION OF THE POSITION AND SHAPE OF CARBON NANOTUBES USING THE AFM

In order to use individual NTs to build nanostructures and devices one must be able to manipulate them and place them at predefined positions at will. The rather strong interaction between the NTs and the substrate on which they are dispersed makes it possible to manipulate their position at room temperature by applying lateral forces of the appropriate magnitude with the tip of an AFM. We have found that the shear stress of NTs on surfaces such as H-surface passivated silicon is high, of the order of 10^7 N/m, such that not only can the position of the nanotubes be controlled but also their shape (Hertel, 1998a).

To manipulate the nanotubes, we have to change the mode of operation of the AFM. While for NT imaging we use the AFM in the *non-contact* mode with very small forces (in the range of pN) applied by the tip, the *contact* mode with normal forces of 10-50 nN is employed for manipulation. In Figure 4.1, we show a sequence of manipulation steps used to fabricate a simple device. Specifically, a single MWNT originally on an insulating substrate (SiO_2) is manipulated in a number of steps (not all shown) onto a tungsten thin film wire (~80 Å high), and, finally, is stretched across an insulating WO_x barrier (itself made by AFM tip-induced oxidation (Avouris, 1997) at an earlier stage). It is interesting to note that highly distorted NT configurations are formed during the manipulation process which, however, are stabilised by the interaction with the substrate. The ability to prepare locally highly strained configurations and the well known dependence of chemical reactivity on bond strain suggest that manipulation may be used to locally strain structures and make them susceptible to local chemistry.

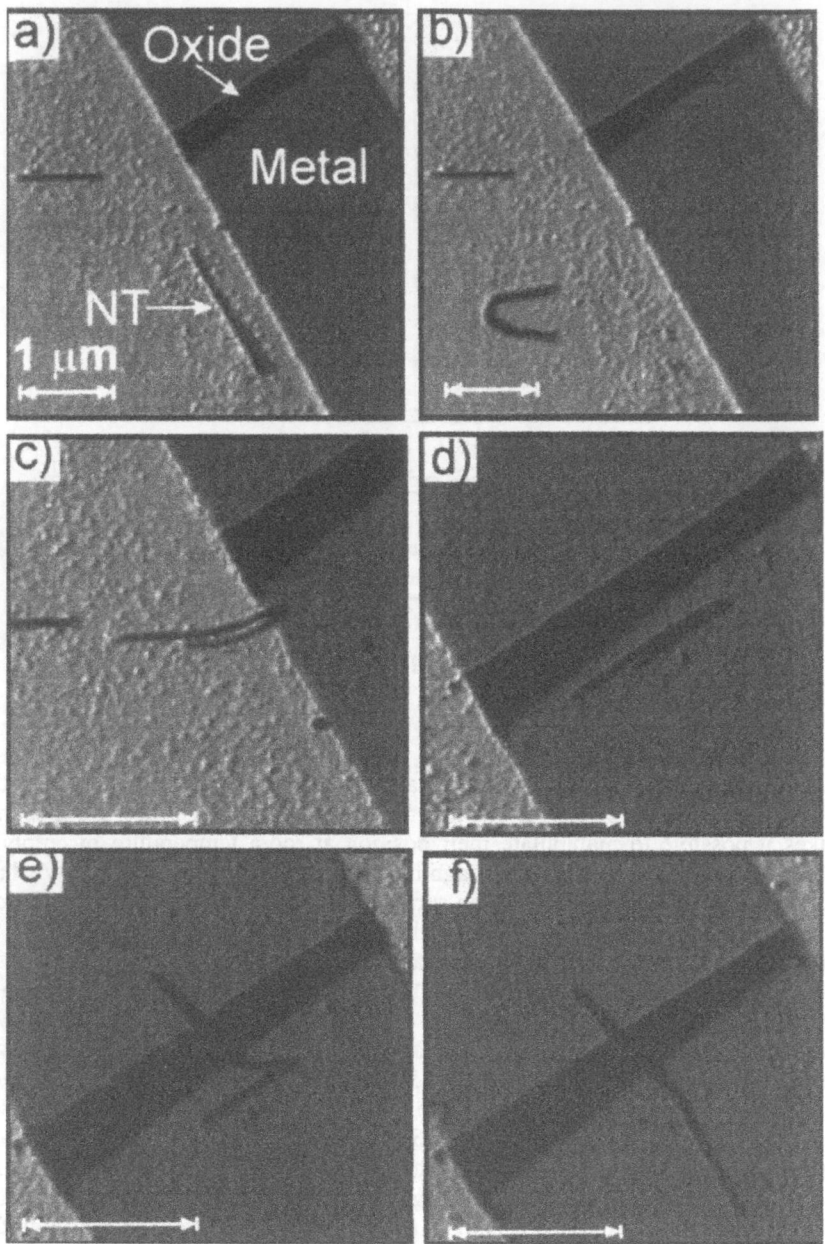

Figure 4.1 AFM manipulation of a single multi-wall nanotube such that electrical transport through it can be studied. Initially, the nanotube is located on the insulating (SiO_2) part of the sample. In a stepwise fashion (not all steps are shown) it is dragged up the 80 Å high metal thin film wire and finally is stretched across the oxide barrier.

5. ELECTRICAL PROPERTIES OF CARBON NANOTUBES

5.1 Some general concepts on transport in low dimensional systems

Before we discuss the process of electrical transport in carbon nanotubes, we review a few concepts important for the description of transport in low-dimensional systems such as the carbon nanotubes. In describing the transport mechanisms, we need to consider the relation of certain characteristic lengths. These involve the elastic mean free path L_m of the carriers and their phase relaxation length L_φ, in relation to the length L, and width W, of the sample. Phase relaxation is caused through inelastic electron collisions with dynamic scatterers such as phonons, magnetic impurities which have a spin that fluctuates with time, or collisions with other electrons. At low temperatures, the dominant mechanism for phase relaxation is provided by electron-electron collisions. Depending on the relation between L_m, L_φ and L we distinguish the following transport regimes in quasi-1-dimensional (Q1D) materials: *classical*, when $L_\varphi < L_m \ll L$; *ballistic*, when $L_m > L_\varphi > L$; *localized*, when $L_m \ll L_\varphi < L$.

In the classical regime, transport obeys Ohm's law with the resistance of a wire being proportional to its width and inversely proportional to its length. In the (strongly) localized regime the resistance of the wire increases exponentially with increasing length (Landauer, 1957, 1992; Anderson, 1980). Of particular interest to us here is the ballistic regime where there is no scattering and transport is coherent. In this case, the resistance is independent of the length of the material. The interest in carbon nanotubes is at least in part due to the possibility of achieving ballistic transport in these materials. The resistance of a ballistic system is determined by the interfaces between the conductor and the leads it is connected to, i.e. it is an interface resistance. Q1D materials like the nanotubes have several sub-bands or modes arising from the lateral confinement of the electrons. The interface resistance of a ballistic system can be viewed as arising from the fact that in the macroscopic leads there are a large number of transverse modes, while the conductor has only a limited number of modes. The Landauer formula provides the basis for the discussion of the conductance of such systems: $G = (2e^2 / h)\sum_i T_i$. In this equation $2e^2 / h$ is the quantum of conductance (77.48×10^{-6} Ω^{-1}) and T_i the transmission probability of mode i, that is the probability that an electron injected from one lead into this mode will emerge from the other end. The sum is carried over all the modes that contribute to the conduction process. When the transport is ballistic, then all transmission probabilities equal 1, and $\sum_i T_i = M$, the number of contributing modes. In the case of NTs at low bias, two transverse modes are expected to contribute to the conductance. Therefore, if the nanotube behaves as a ballistic 1D-system its conductance should be $4e^2 / h$, which corresponds to a resistance of ~6 kΩ.

There are two properties of a ballistic system that have important implications for the use of nanotubes in nanoelectronic devices: (a) speed of signal propagation, and (b) low energy dissipation. In particular, it is important to note that in the case of ballistic transport, the energy is not dissipated inside the nanotube, but at the lead-nanotube interface where cooling is most efficient. This fact should further contribute to the stability of nanotubes when used as interconnects.

In most experimental situations, transport may not be truly ballistic but instead may involve a number of elastic collisions ($L_m < L$). Depending on the importance of these collisions we talk about *quasi-ballistic* or *diffusive* transport. As we discussed above, elastic scattering from impurities does not destroy the phase of the electron wavefunction. The coherent motion of the electrons allows quantum interference phenomena to take place; in particular it may lead to the phenomena of *coherent backscattering, weak localization,* and *universal conductance fluctuations* in disordered systems at low temperatures (Datta, 1995; Bergman, 1984; Beenakker, 1991; Washburn, 1986)

Coherent back-scattering and weak localization can be understood by considering the interference of the Feynman paths from one trajectory point of the electron (or hole) to another. Of particular importance are electron trajectories that close upon themselves, so that the particle returns to its starting point (see Fig. 5.1a). An electron wave can traverse such paths by moving either clockwise or anti-clockwise around the loop. Because of time-reversal invariance, the phase change acquired moving in the two opposite directions is the same, therefore, the two waves can interfere constructively at the origin leading to enhanced back-scattering. The loops of coherent back-scattering form standing waves that do not carry any current. The electrons are effectively stuck in the loop, their ability to diffuse through the sample is reduced, and momentum relaxation, i.e. the resistance, is increased above its classical value. This is the basis of the weak localization effect.

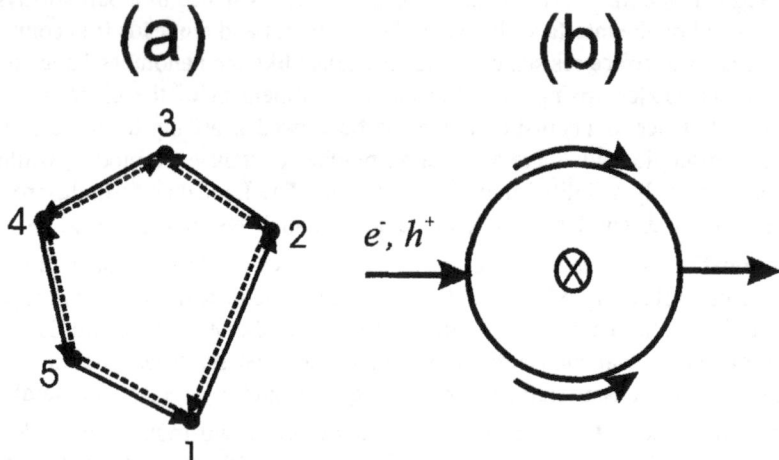

Figure 5.1 (a) An elastic scattering path that closes upon itself leading to coherent back-scattering.
(b) The geometry of the Aharonov-Bohm effect.

One way of eliminating coherent back-scattering and suppressing weak localization is by applying an external magnetic field. In the presence of the field, the phase change acquired by an electron wave propagating along a path *l* is given by the line integral of the vector potential **A** along this path:

$$\delta_l = \exp\left[\frac{ie}{\hbar}\int_l \mathbf{A}\cdot d\mathbf{l}\right].$$

For a closed loop the two counter-propagating waves will now acquire opposite phases and the constructive interference leading to coherent back-scattering will be destroyed. By invoking Stokes' theorem we can write $\oint A \cdot dl = BS$. This shows that the phase change is proportional to the magnetic flux passing through the area that is enclosed by the scattering loop. Every time an extra flux quantum $\Phi_0 = h/e$ enters the loop, the relative phase changes by 2π. Since the area enclosed by the two counter-propagating electron waves is twice the area of the loop, the period of the interference should be $h/2e$.

An analogous situation is encountered in the *Aharonov-Bohm* effect, where one measures transport through a very small and thin metallic ring in the presence of a perpendicular magnetic field (see Fig. 5.1b). The current exiting the ring is modulated due to alternating constructive and destructive interference (Washburn, 1986). In this case the interference is measured at the opposite site from the entrance of the ring and, therefore, a cycle is completed when the flux is h/e. NT rings may reveal such Aharonov-Bohm oscillations.

5.2. Experimental studies

There have been a number of studies of the transport properties of carbon nanotubes. In general, two probe measurements of single MWNTs at 300 K show resistances of the order of a few tens of kΩ, while SWNTs show much larger resistances of the order of 1 MΩ. It is generally believed that these high values reflect the resistance of the tube-metal electrode (usually Au or Pt) contacts. Early four-probe measurements gave resistivities that span four decades (Ebessen, 1996). This large spread reflects party the presence of both metallic and semiconducting tubes, and partly the presence of variable numbers of defects in the tubes. These defects may have been in the original nanotube material, or were introduced during the in situ fabrication of the electrodes using ion beams. The influence of defects on the resistivity was demonstrated by the AFM studies of Dai et al. (Dai, 1996).

For electrical applications, it is of great interest to know how much current can be passed safely through an individual nanotube. A coupling between the electrons and the vibrations of the nanotube would lead to energy dissipation, thus raising the temperature of the tube. For high enough currents the tube will be destroyed. In Figure 5.2a we show a single MWNT bridging two gold electrodes. The current through this tube was increased until the tube broke (Fig. 5.2b). As seen in Fig. 5.2c, this MWNT was able to carry a current of up to ~60 μA or a current density of the order of 10^{11}-10^{12} A/m^2. Such current densities are much higher than those that can be supported by metal wires of similar size. This is the result of low energy dissipation rates and of the high stability of NTs.

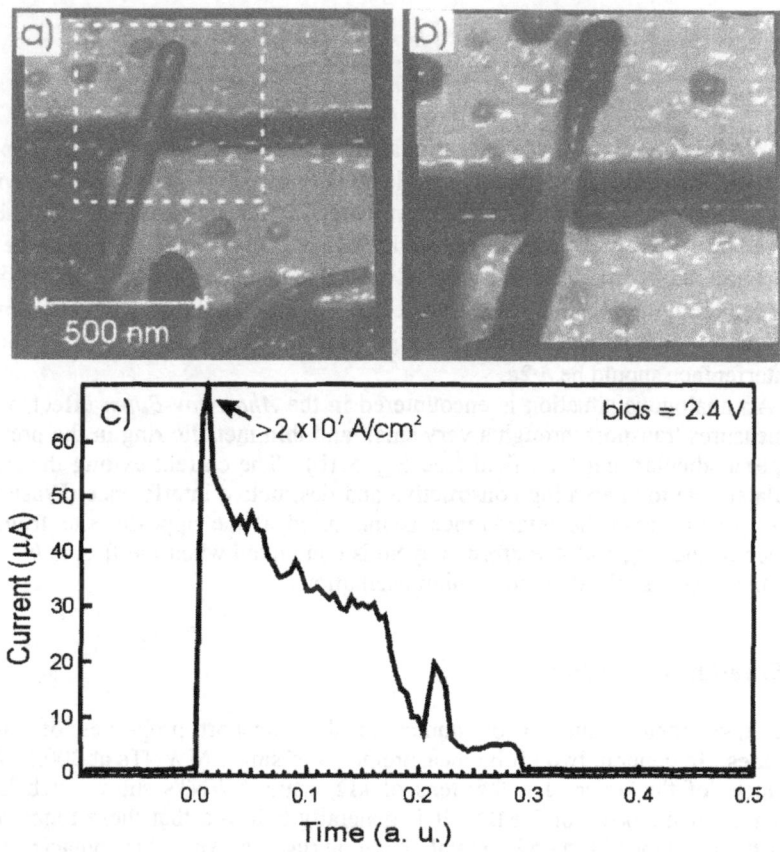

Figure 5.2 (a) Multi-wall nanotube bridging gold electrodes. (b) The nanotube after electrical breakdown. (c) Time-dependence of the current during breakdown.

Several authors have measured the temperature dependence of the resistivity of a variety of different NT samples. For example, ropes of SWNTs gave a 300 K resistivity of $\rho = 0.03$-0.1 mΩcm with a positive $d\rho / dT$ above T=35 K which changed to negative at lower T (Fisher, 1997). The actual resistivity value of ropes was found to depend on the purification and annealing procedures to which the ropes were exposed. Measurements on an individual SWNT, on the other hand, showed that charging and Coulomb blockade develops at low temperatures due to the high contact resistance (Tans, 1997). A way by which the contact resistance can be reduced was demonstrated by Bachtold et al. (1998a). It involves irradiation of the contacts with high-energy electrons in the SEM. Our group has made similar observations (Shea, 1998a). Very recently, Bachtold et al. (1998b) managed to fabricate ohmic contacts and perform four-probe measurements on 9.5 nm diameter MWNTs. By depositing gold electrodes on top of the MWNTs, contact resistances to individual MWNTs as low as 300 Ω were observed. These authors also studied the magneto-resistance of the system and concluded that their MWNTs behave like a weakly localized two-dimensional system, with a coherence length $L_\varphi > 200$nm at T=4.2 K. Similar conclusions were reached by Langer et al. (1996) who found an $L_\varphi \sim 10$ nm. Frank et al. (1998) has provided evidence for quantized conductance in MWNTs. These authors used MWNTs attached with one end to the tip of an STM, while the other was immersed in mercury or a liquid metal. They found that the conductance was one unit of the conductance quantum, i.e. $G_0 = (12,900 \ \Omega \ )^{-1}$. However, we should note that the expectation for the low bias conductance of an NT is $2G_0$.

Little is known about transport in SWNTs. At low T, charging and Coulomb blockade dominate due to the high contact resistances. However, Tans et al. (1997) used energy quantization arguments to analyse such Coulomb blockade experiments and inferred an $L_\varphi \sim 3$ μm for a SWNT at 4.2 K.

Linear SWNTs are very close to ideal 1D-systems. Because of the lack of self-folding trajectories in such systems, the powerful technique of magneto-resistance cannot be applied directly to study their transport properties. However, SWNT rings have the required geometry (Fig. 5.3) and thus allow us to investigate the role of electron coherence at low temperature. The results are very recent and not yet fully analysed. Here, we present some of the results as an example of possible new directions and challenges in nanotube research.

There have been already some theoretical and experimental studies of the magneto-resistance of linear carbon nanotubes. Seri and Ando (1997) examined theoretically the Boltzmann conductivity of NTs. They predicted that undoped metallic nanotubes will have a large positive magneto-resistance due to the combined effects of the reduction in electron velocity and the shrinkage of the wave function for states $k\sim 0$. For heavily doped NTs they predicted a large negative magneto-resistance. Langer et al. (1996) examined the magneto-resistance of large diameter individual MWNTs (Such MWNTs are effectively 2D-systems). They observed a negative magneto-resistance, which they interpreted as resulting from the removal of the weak localization by the magnetic field. They also observed additional structure at higher fields, which they interpreted in terms of *universal conductance fluctuations*. A negative mageto-resistance was also observed by Baumgartner et al. (1997) using mats of aligned MWNTs.

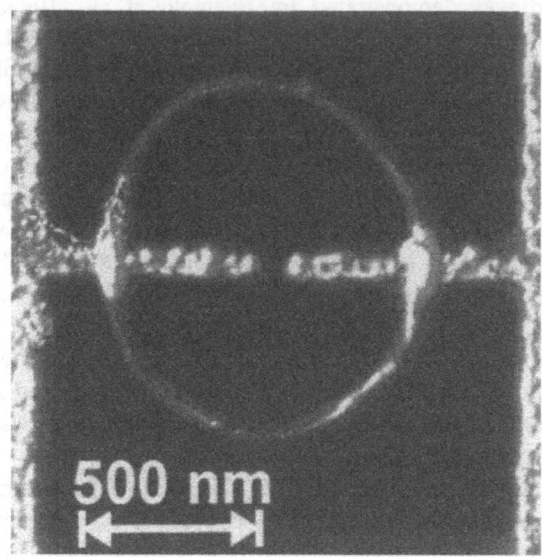

Figure 5.3 Nanotube ring on gold electrodes.

As we already mentioned, SWNT rings allow one to examine the effect of a magnetic flux penetrating the ring on the transport properties, and investigate the role of electron coherence at low temperatures. In Figures 5.4 and 5.5 we present some results from our initial studies of these systems (Shea, 1998b). Figure 5.4 shows the dependence of the differential resistance of a 0.42 μm radius ring at 4 K (contact resistance: ~20kΩ) as a function of the strength of a magnetic field perpendicular to the plane of the ring, for three values of the excitation current. The resistance is seen to decrease smoothly with increasing magnetic field. In the range of field strength studied (0-5T), the effect amounts to an 15% reduction of the resistance. Furthermore, the value of the resistance at $H = 0$ is found to increase with decreasing current. The fact that biasing the back gate of the device has no effect on the measurements and the lack of the corresponding structure in the I-V curves themselves, suggest that the results are not influenced by Coulomb blockade. As we discussed above, the negative magneto-resistance observed is in accord with expectations based on the destruction of weak localization by the magnetic field. At the low temperature of the experiment, electron-electron scattering should be the dominant dephasing mechanism; therefore, the dependence on the current can also be understood in terms of the effects of dephasing on weak localization (10pA, 1nA, 100nA correspond to electron temperatures of 10mK, 1K and 70K, respectively). In Figure 5.5 we show the low temperature dependence of the magneto-resistance. The 4.2 K and 2.5 K results are consistent with the trend expected from weak localization, an increase in the resistance with increasing T and a weak negative magneto-resistance. At 0.4 K (10pA, T_{el}~10mK), however, there is a dramatic change. There is a large increase in the resistance, and the effect of the magnetic field is greatly enhanced. A 1T field is seen to reduce the resistance of the ring by ~50%. Additional non-periodic fluctuations are observed at higher fields. They are due to universal conductance fluctuations (Datta, 1995; Beenakker, 1991). By fitting the dV/dI vs. H data to the predictions of weak 1D localization theory we can extract the coherence length of the electrons. At 3 K this length reaches 520 nm.

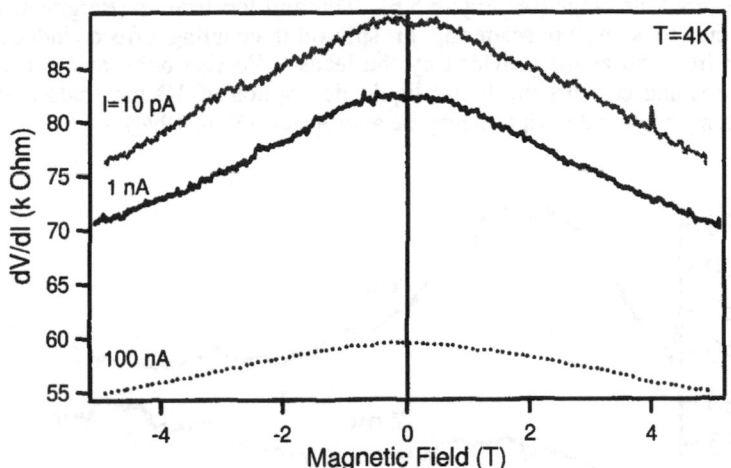

Figure 5.4 Differential magneto-resistance of a SWNT ring (radius = 0.42 μm) taken at 4K using different current biases. The magnetic field applied is perpendicular to the plane of the ring.

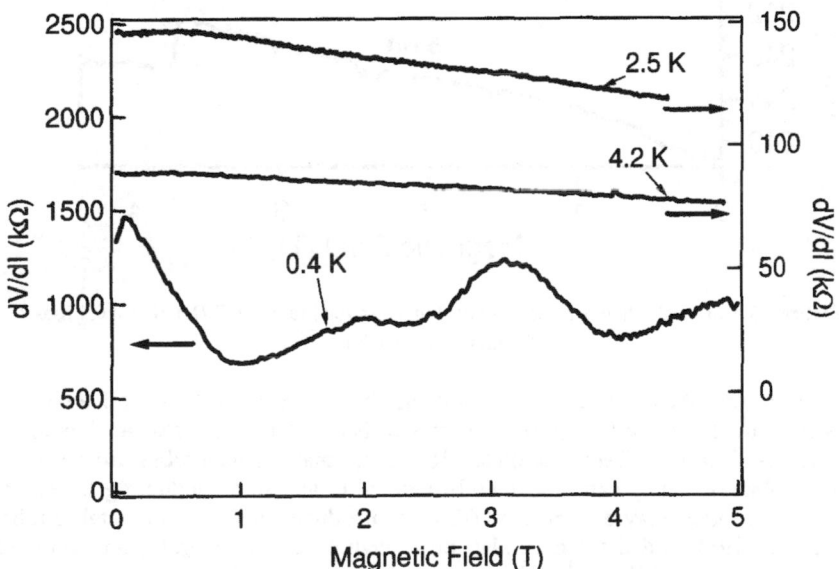

Figure 5.5 Influence of the temperature on the differential magneto-resistance of a SWNT ring.

At the lowest temperatures of our experiments (0.3-0.4K) we sometimes observed a weak positive magneto-resistance at low fields which turned into negative magneto-resistance at higher fields. Furthermore, in an experiment where the contact regions were irradiated by high energy electrons to reduce the contact resistance, we observed a positive magneto-resistance (anti-localization behavior) over the entire field range (see Figure 5.6). This anti-localization (Bergman, 1984) may be due to spin flip scattering or spin-orbit coupling effects induced by scattering from Au atoms provided by the leads. We also point out that at low temperatures and currents the Fermi liquid description of 1D materials may fail and a Luttinger liquid description may be appropriate (Voit, 1994).

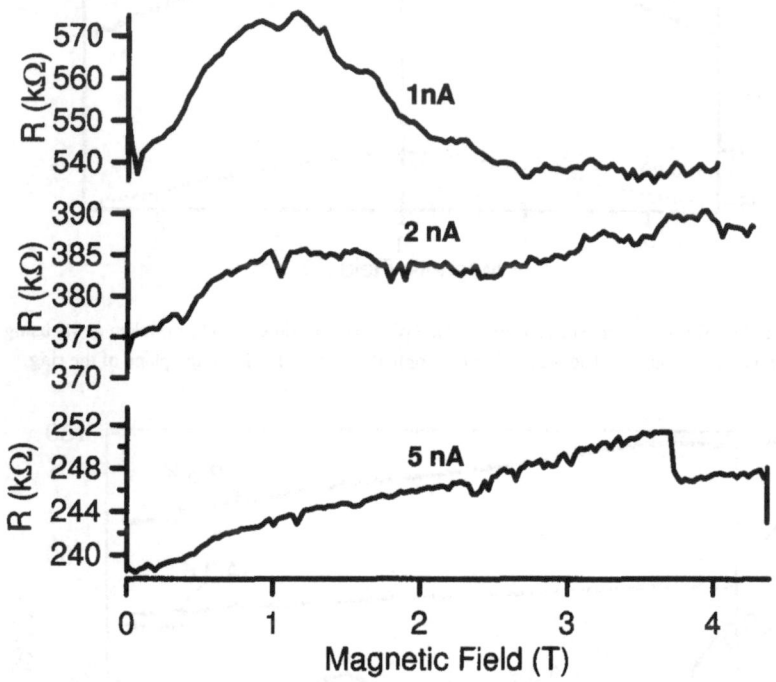

Figure 5.6 Anti-localization behavior observed in a electron irradiated SWNT ring at 0.4 K as a function of current bias.

From the above discussion, it is clear that there are still several answered questions involving the transport properties of NTs. Part of the reasons lies in the samples used in the different studies. Ropes and mats of nanotubes are not well-defined objects. They can be a blend of metallic and semiconducting tubes, the aggregation forces may deform the NTs and introduce band-gaps in metallic tubes (Delaney, 1998), and there may also be inter-tube electrical transport involved. Furthermore, single MW and SW tubes may have defects of different types. These may be point defects such as vacancies and substitutional defects, or structural deformations. Calculations by Chico et al. (1996) of the influence of defects have shown that a single vacancy in a (4,4) NT can lead to a 50% reduction of its conductance (the effect decreases as the diameter of the NT is increased.) One of the most important bottlenecks in the understanding and electrical applications of NTs involves the elimination of the high contact resistance.

6. THE NATURE OF CONTACT RESISTANCE AND EFFECT OF STRUCTURAL DISTORTIONS ON ELECTRICAL TRANSPORT

A different type of NT defect is provided by structural deformations. As we discussed in section IIA, due to their strong adhesion to the substrate, NTs tend to deform so as to follow the substrate topography and thus optimize their adhesion. A particularly relevant distortion arises when NTs are draped over metal electrodes in electrical transport studies. This process can bend and strain the NTs to a variable degree. We note that all the carriers in NTs are located on their surface and are involved in bonding, a situation very different from that of conventional metal wires. One would expect that changes in bonding and also surface interactions might affect transport significantly. To investigate this problem, we undertook in collaboration with Dr. Alain Rochefort and his colleagues at the Centre de Research en Calcul Applique (CERCA) a study of the electronic structure and transport properties of deformed (bent) SWNTs. The calculations involve the optimization of the geometric structure of bent nanotubes using molecular mechanics (see Fig. 6.1) with the optimized geometry then being used as input to the electronic structure programs. A key new element in these calculations is the inclusion of both σ and π-electrons. Previous calculations on bent tubes have included only the π-electrons and concluded that the effects are negligible (Kane, 1997).

The hybridization of a carbon atom in graphite is sp^2 with the resulting three hybrids forming σ-bonds with three neighboring C-atoms, while the fourth electron occupies a p-level and participates into the π-bonds. It is these $C2p_\pi$-electrons that participate in the electrical conduction process. It is important to note that the σ and π electron densities are orthogonal so that, to a first order, their interaction can be neglected or taken into account in the form of an effective potential. The sp^2 hybridization is more or less preserved on going from a graphene sheet to a nanotube, provided that the induced curvature is small. Upon bending, however, the curvature can be very high and one would expect that this will induce the mixing of σ and π-states which are not orthogonal anymore. The mixing of the π-states with the higher binding energy σ-states will push the former up to lower binding energies and introduce σ-character in them (a re-hybridization process). Thus, while in a straight NT the conduction electrons have essentially pure $C(2p_\pi)$ character, they acquire significant $C(2p_\sigma)$ character upon bending.

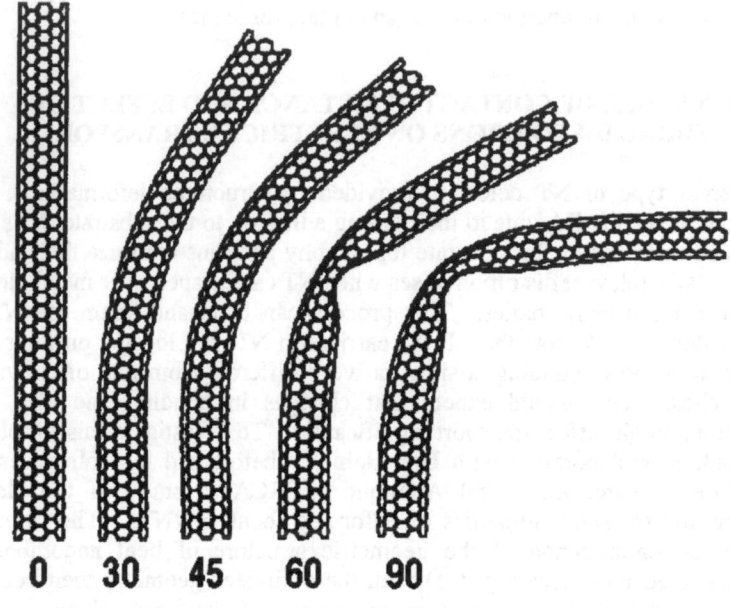

Figure 6.1 Structures of isolated bent nanotubes optimised by molecular mechanics and used in the electronic structure computation. The bending angles are given in degrees.

In the calculations, we utilized finite segments of carbon nanotubes and computed the electronic structure with the extended Hückel method. In Fig. 6.2 we show results for a (6,6) nanotube segment involving 948 carbon atoms (Rochefort, 1998a). The quantity shown is the local density of (π) states (LDOS) in the neighbourhood of E_F. The position of the circular segments whose LDOS is shown is indicated by a number which runs from 1 (end of the tube) to 40 (center of the tube). In straight tubes there is no significant difference in the LDOS at the end and in the middle of the tube (some small differences due to finite size of the segment are present). As the bending angle increases, however, so do the changes in the LDOS. The change is particularly strong at $60°$ and $90°$ bending angles where the strain is high enough to induce a kink in the NT structure. As we expected there are new occupied states appearing near E_F. An analysis of their composition indeed shows the expected increased $C(2p_\sigma)$ character resulting from $\sigma-\pi$ mixing (Rochefort, 1998a).

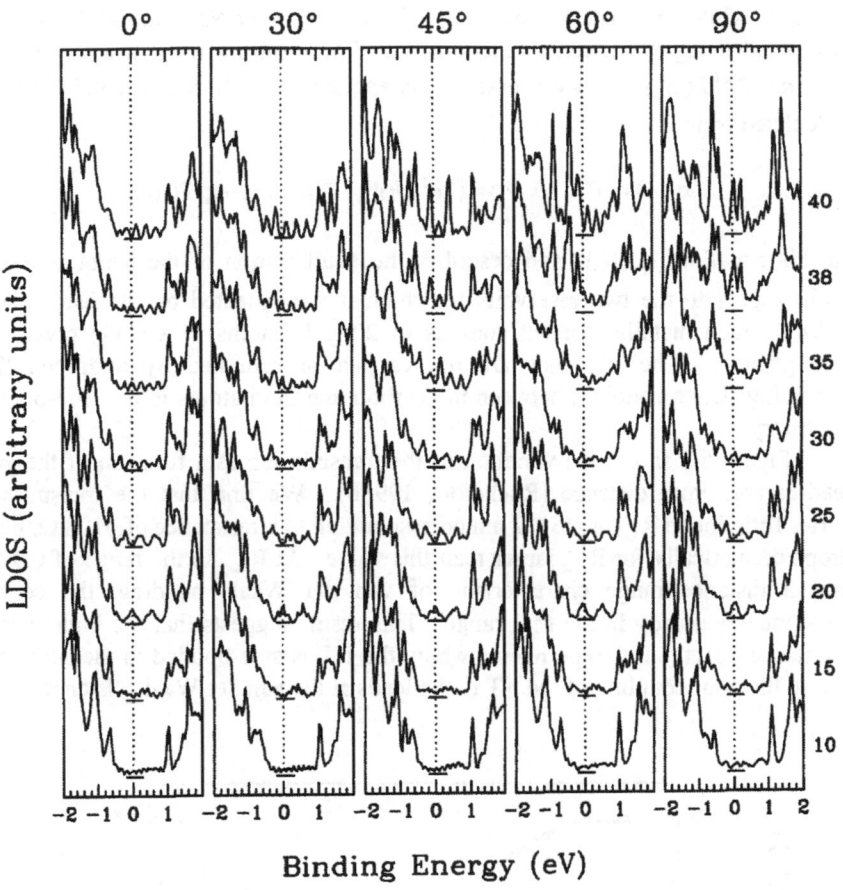

Figure 6.2 Variation of DOS and LDOS near the Fermi level for several bent (6,6) nanotubes. Indices give the relative position of the section in the nanotube structure.

To find what effect this local bending-induced re-hybridization may have on the transport properties of the NTs, we first need to couple the nanotubes to macroscopic metal leads and compute the effect of these leads on the electronic structure of the NT. The electrical transport properties of the combined system are described in terms of the retarded Green's function (Datta, 1995; Economou, 1983), and are computed using the Landauer-Buttiker formalism as applied in a recent paper by Tian et al. (1998). The key element of this approach is the description of the infinite leads by self-energies. The transmission T(E) is then calculated from the Green's function: $G_{NT} = [ES_{NT} - H_{NT} - \Sigma_1 - \Sigma_2]^{-1}$, where S_{NT} and H_{NT} are, respectively, the overlap and Hamiltonian matrices, and Σ_1 and Σ_2 are the self-energies accounting for the effects of the two contacts. The Hamiltonian and overlap matrices are computed using the extended Hückel method, while the transmission T(E) of the *gold lead-NT-gold lead* system is computed using the Green's function G_{NT}. The resulting resistance is given by $R = h/(2e^2 T(E))$. When a potential is applied, the differential conductance is calculated from:

$$\sigma(V) = (\partial\ I / \partial\ V) \approx (2e^2 / h)[\eta\ T(\mu_1) + (1-\eta)T(\mu_2)]$$

In this equation, $\eta \in [0,1]$ and describes the equilibration of the nanotube levels with respect to the two reservoirs to which it is connected by the leads (Tian, 1998). Each metallic contact consists of 22 gold atoms in a (111) crystalline arrangement. The NT ends are not capped or saturated by hydrogen thus providing strong coupling between the carbon and gold atoms in this end-bonded geometry.

Figure 6.3 shows the variation of the transmission as a function of the gold lead-carbon atom distance (Rochefort, 1998b). We find that the transmission varies little for an R_{Au-C} up to 0.2 nm corresponding to a resistance of ~10 kΩ, but it drops dramatically for R_{Au-C} larger than this value. At R_{Au-C} in the range of 0.3-0.4 nm, a distance range characteristic of van der Waals bonding, the contact resistance is already in the MΩ range. This result suggests that the high contact resistance observed in experiments where the NT is side-bonded to the surface is due to the poor coupling of the NT to the leads at the van der Waals distance.

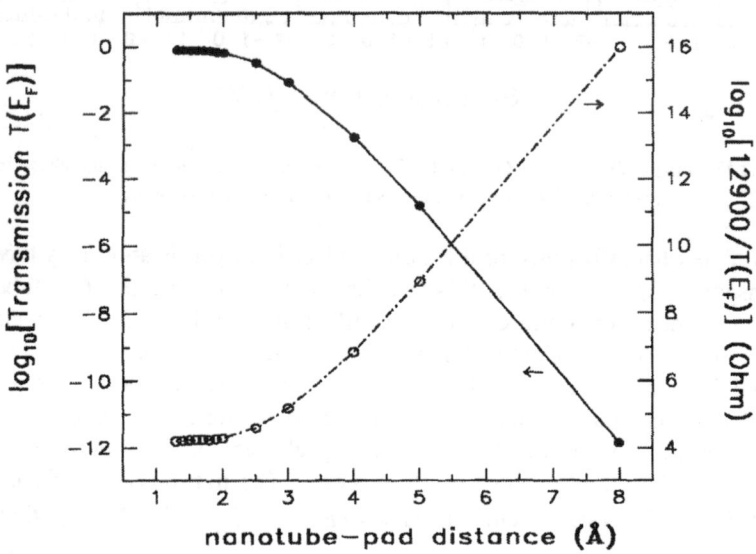

Figure 6.3 Transmission T at E_F as a function of the nanotube-gold pad distance.

Next we consider the change in resistance due to bending. Figure 6.4 shows the differential resistance as a function of applied bias for two extreme cases of equilibration of the Fermi levels (Rochefort, 1998b); the first is when $\eta = 0$ (Figure 6.4a), and the symmetric case $\eta = 0.5$ (Figure 6.4b). When $\eta = 0$ (i.e. the Fermi level of the nanotube follows exactly the applied voltage on one gold pad) the conductance is directly proportional to the transmission function. The differential resistance shows little change upon bending by 30° or 45°, but a large increase, by almost an order of magnitude, is found at 0.6 V for NTs bent by 60° or 90°. Similar effects are observed when $\eta = 0.5$ (i.e. when the Fermi level of the nanotube is floating at half the voltage applied between the two gold pads) with the non-linear resistance of the 60° bent NT increasing by about a factor of 4 from the computed resistance of the straight NT. These results suggest that there is a critical bending angle (between 45° and 60°) above which the conduction in carbon nanotubes is drastically altered. Differential resistance peaks localised in certain energy ranges as a result of re-hybridisation are a new finding, but we expect that they should appear in other covalently-bonded molecular wires.

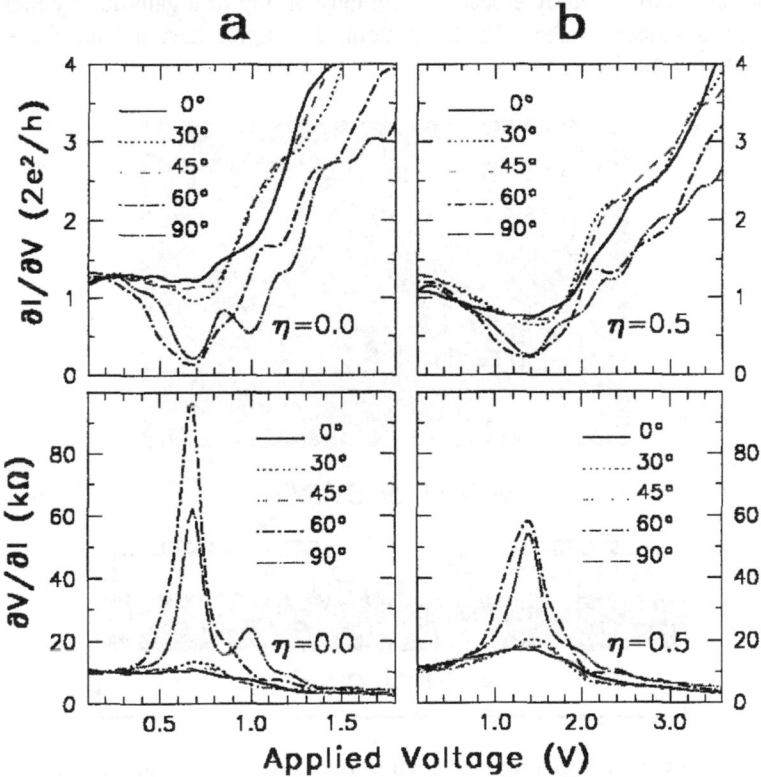

Figure 6.4 Differential conductance (top) and resistance (bottom) of bent tubes for two extreme cases where (**a**) $\eta = 0$ and (**b**) $\eta = 0.5$.

7. CARBON NANOTUBE DEVICES: FIELD-EFFECT AND SINGLE ELECTRON TRANSISTORS

7.1 Field effect nanotube devices

As we saw above, metallic NTs are excellent conductors. Even more interesting applications can be envisioned for semiconducting SWNTs, which can be used as active elements of electronic devices, in particular, field-effect transistors. Ultra-small size, room temperature operation, no need for doping, and a conventional switching mechanism would be the highly desirable characteristics of such a device. Recently, two groups (Tans, 1998; Martel, 1998b) succeeded in using a semiconducting SWNT as the channel of a field-effect transistor (NT-FET).

Figure 7.1 shows two simple configurations of an NT-FET. In the top design the NT is positioned so as to bridge two gold electrodes, which act as the *source* and *drain* of the FET, while the middle electrode is used as *side gate*. The Au electrodes are fabricated by e-beam lithography on top of a gate-quality SiO_2 film grown on a silicon wafer. In the bottom, a simpler design uses the silicon waferitself as a *back gate*.

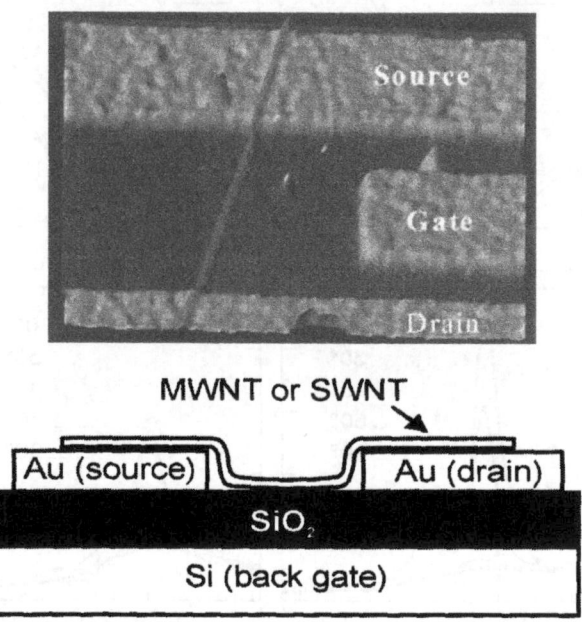

Figure 7.1 Two configurations of field-effect transistors based on carbon nanotubes.

In Fig. 7.2a we show the output characteristics $I\text{-}V_{SD}$ of one of our back-gated NT-FET devices consisting of a single SWNT with a diameter of ~1.6 nm, as a function of the gate voltage at room temperature. At $V_G = 0$, the $I\text{-}V_{SD}$ curves are linear and the resistance is R=2.9 MΩ. For V$_G$<0, the I-V_{SD} curves remain linear, whereas they become increasingly non-linear for V_G >>0 V up to a point where the current becomes unmeasurably small, indicating a controllable transition between a metallic-like state (linear I-V_{SD} curves) of the NT and a semiconductor state (non-linear I-V_{SD} curves).

The transfer characteristics of the same device, i.e. the variation of the source-drain current I as a function of the gate voltage V_G for several source-drain biases V_{SD}, are given in Figure 7.2b. It is clear that the gate can strongly modify the current flow through the nanotube. The enhancement of the current at negative gate bias suggests that positive holes are the main carriers. Figure 7.2 also shows that this NT-FET is a normally "on" device which can be switched to the "off" state by a positive gate bias. As the inset in Figure 7.2 shows, the gate can modulate the low-bias conductance $G = I/V_{SD}$ of the NT by 5 orders of magnitude. At V_G<0 V, the I-V_G curves saturate, indicating that the contact resistance R_C at the metal electrodes starts to dominate the total resistance R=R_{NT}+ R_C of the device. The saturation value corresponds to $R_C \approx$1.1 MΩ. It is clear that if we were to decrease R_C, then we could enhance the effect of the gate.

If we assume that the carriers are rather uniformly distributed along the length of the NT, as is suggested by the fact that the I_{SD} current is non-zero at V$_G$ = 0, we can estimate the total charge on the NT. By writing the total charge as $Q_T = CV_{G,T}$, where C is the NT capacitance and $V_{G,T}$ the threshold voltage needed to deplete the NT, we obtain for the NT in Fig. 7.2 a linear carrier density of $p = Q / eL \approx$ 9x10^6 cm^{-1}. This large current density represents the sum of the intrinsic carriers and the carriers (holes) that result from the NT-to-gold pad electron transfer.

Multi-wall nanotubes tend, in general, to have a large diameter. Since the band gap of semiconducting NTs decreases with increasing diameter (Mintmire, 1992), i.e. $E_{GAP} = 1 / d$, nanotubes with d > 12 nm should be effectively metallic at 300 K. Indeed, we fabricated a number of MWNT devices with R~ 100 kΩ and found no gate action (see Fig. 7.3, curve A). However, as we already pointed out, structural distortions alter the electronic structure of an NT. In Fig. 7.3 (curve B) we show the behaviour of a collapsed MWNT. As in the case of the SWNT-FET, the source-drain current decreases with increasing gate voltage. In contrast to the SWNT-FET, however, this MWNT-FET cannot be completely depleted. Between V$_G$ = -35 and 25 V the resistance increases only by a factor of 2. A possible explanation of this behaviour invokes two current-carrying channels provided by the bottom and the top outer layers of the collapsed NT (see Fig. 7.4). The bottom channel is responsible for half of the current and can easily be switched off by the field of the gate. The top channel, however, is partially screened by the bottom layer (when it is conducting) and the other intervening layers and is not effectively switched off.

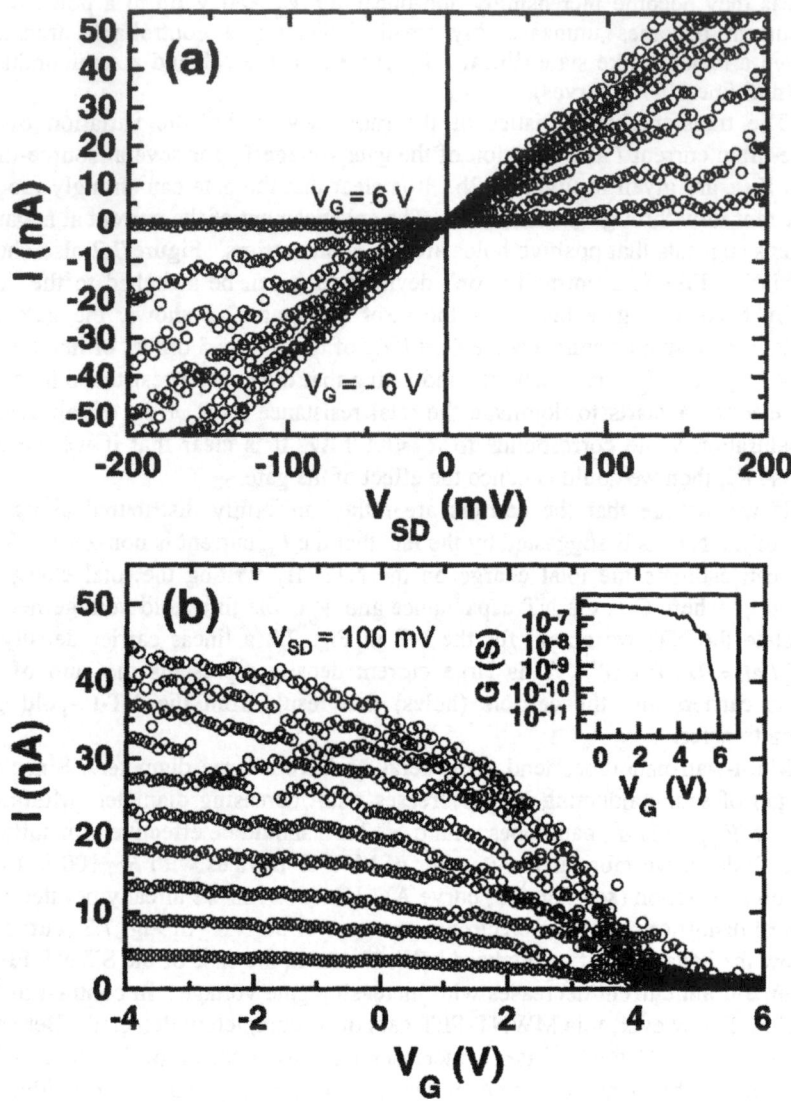

Figure 7.2 Field effect transistor based on a single SWNT. (a) Source-drain current as a function of source-drain biases V_{SD} measured for different gate voltage (V_G = -6, 0, 1, 2, 3, 4, 5, and 6V). (b) Source-drain current I as a function of gate voltage V_G for several source-drain biases V_{SD}. The insert shows the variation of the low-bias conductance G= I/V_{SD} as a function of the gate voltage.

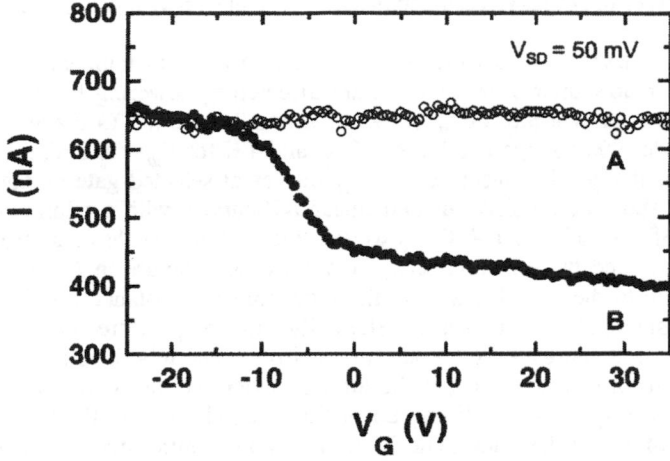

Figure 7.3 I-V$_G$ curve of a typical MWNT device (curve A), and that of a collapsed MWNT of similar cross section (curve B).

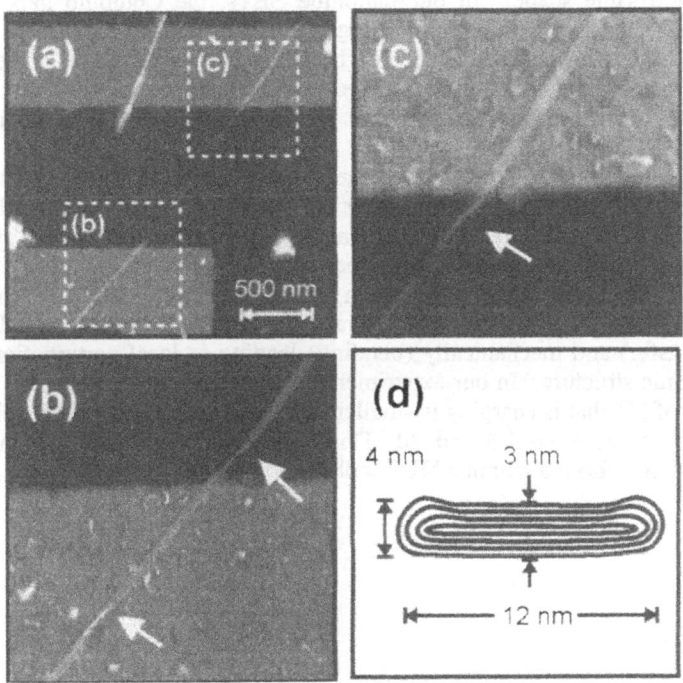

Figure 7.4 (a) AFM picture of the MWNT-FET. **(b)** and **(c)** Close up views showing three twists in the collapsed NT (see arrows). **(d)** Schematic cross section of the collapsed MWNT.

7.2 Field-effect to single-electron transistor conversion at low temperatures

The NT-FET devices we discussed above operate at room temperature. Let us now consider how their characteristics are affected by lowering the temperature. In Fig. 7.5 we show a plot of I_{SD} vs. V_G of a bundle of SWNTs draped over two gold electrodes 400 nm apart at 290 K, 77 K, and 4 K for V_{SD} = 20 mV, 50 mV, and 5 mV, respectively. The insets are I_{SD}-V_{SD} curves at selected gate voltages. The 290 K data show FET behaviour and linear I-V curves, with a minimum device resistance of 250 kΩ. At 4 K, however, the I-V curves have a pronounced Coulomb gap, and there is very sharp non-periodic structure in I_{SD} vs. V_G. It is clear that due to their small size and the high contact resistance the FET device turns into a single electron transistor (SET) when the temperature is low enough so that the charging energy $e^2/2C$ to add an electron to the tube is much larger than $k_B T$. The contact resistance acts as the tunnel barrier and the sharp lines in the I_{SD} vs. V_G plot correspond to the lining up of the internal states in the tube with the Fermi level of one of the leads. The 77 K data show features due to both FET and Coulomb blockade behaviour.

Data at 4 K from most of our SWNT devices show structure corresponding to the addition of one electron at a time to the tube. Each diamond in the gray scale dI/dV_{SD} vs. V_G and V_{SD} plots, as shown in Fig. 7.6a, corresponds the change the number of electrons on the tube by one. For a single metallic island, the diamonds all have the same shape. In our nanotube SETs, the Coulomb gap oscillates aperiodically with V_G, suggesting single-electron transport through multiple Coulomb islands formed within the NT bundle. The depletion of carriers in the tube, and the associated decrease in tube capacitance, are clearly visible in Fig. 7.6b as an increase in the width of the blockade region for large positive gate voltage.

One can determine the charging energy of a length of NT by calculating its capacitance. Alternatively, one can infer the length of the part of the tube that is charging from the measured charging energy. It has been observed by several groups that only the section of a nanotube contained between the metal electrodes acts as the Coulomb island (Bezryadin, 1998, Bockrath, 1997). This is not surprising, given that the electrodes interact with the nanotube chemically (i.e by charge-transfer) and mechanically (bending) leading to local perturbations in the NT electronic structure. In our experiments with bundles of SWNTs we find that the length of NT that is charging is smaller than the spacing between the electrodes by a factor of between 1.5 and 20. This behaviour is probably attributable to multiple islands formed within a NT bundle.

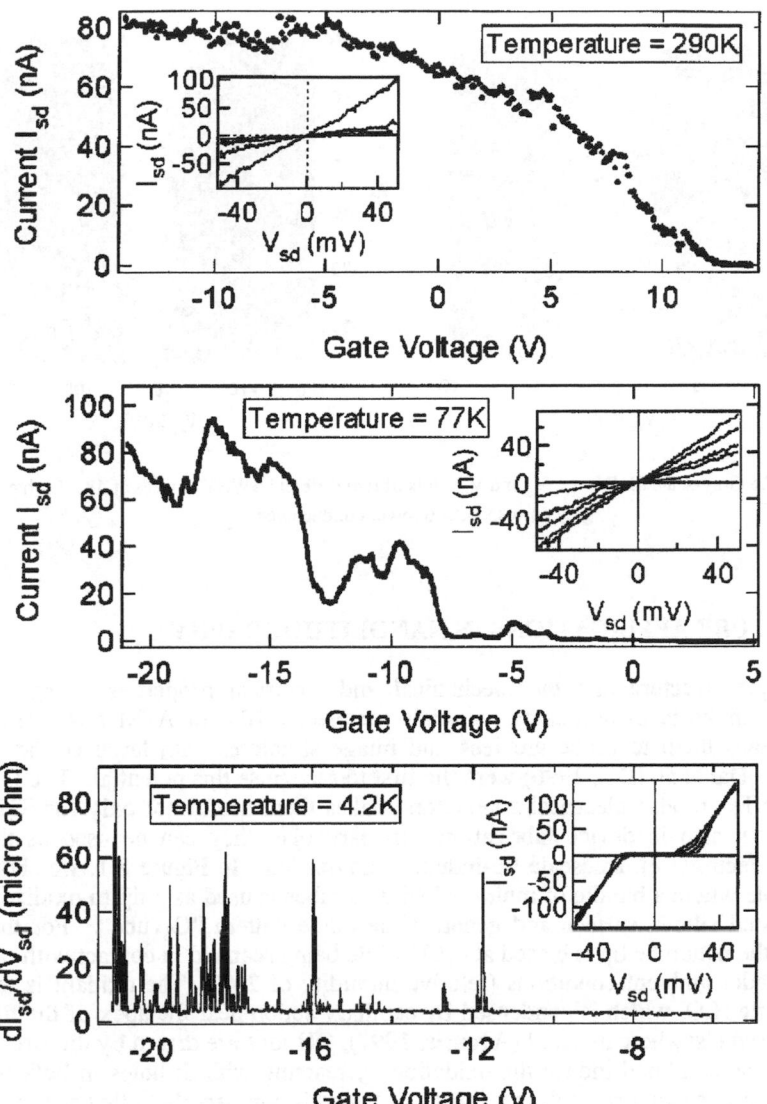

Figure 7.5 Source-drain current as a function of gate bias V_G of a back-gated device composed of a bundle of SWNTs at 290K, 77K and 4.2K. The device behaves like a FET at 290K, and as a SET at 4.2K.

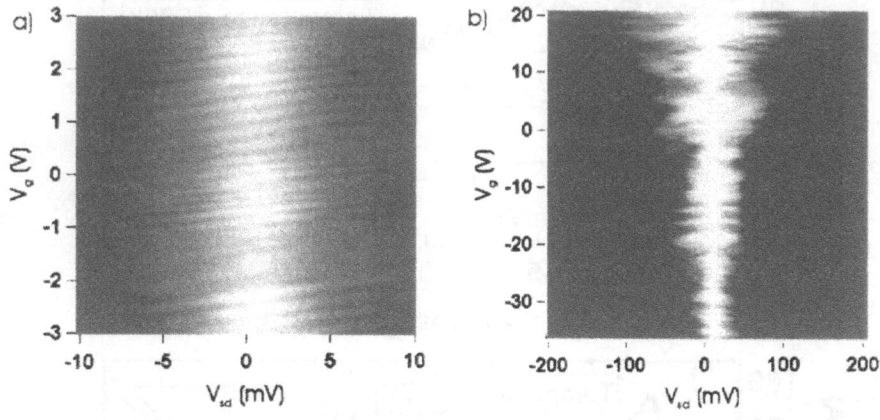

Figure 7.6 Gray-scale dI/dV_{SD} vs. V_G and V_{SD} plots of two different SWNT devices at 4K. Lighter grey corresponds to lower conductance.

8 NANOTUBE APPLICATIONS IN NANOLITHOGRAPHY

The unique structure and the mechanical and electrical properties of carbon nanotubes make them promising materials for use as STM or AFM tips. Their shape allows them to probe crevices and image structures with large curvature gradients. Dai et al. (Dai, 1996) were the first to recognise this potential. The fact that the NTs are also electrical conductors makes them useful not only for STM imaging but also in device fabrication. In particular, they can be used as the negative electrode in nanoscale tip-induced anodization. In Figure 8.1, we show an example where a bundle of multi-walled nanotubes is used as a tip to oxidize a H-passivated silicon surface and generate the oxide pattern "C-Tube". For this purpose, the nanotube tip is biased at -10V while being scanned in contact with the surface under ambient conditions (relative humidity of 20%). The oxidant is the atmospheric H_2O, which is condensed by capillary action near the apex of the tip. As discussed elsewhere in detail (Avouris, 1997), OH⁻ ions are driven by the strong field into the solid and induce the oxidation by reacting with Si holes in bulk Si. Although in this experiment the resolution obtained is comparable to that achieved with conventional tips, we believe that it can be enhanced significantly by using single-nanotube tips. Similar results were reported recently by Dai et al. (1998).

Figure 8.1 Use of conducting carbon nanotube AFM tips as nanofabrication tools. In this case, a nanotube-bundle tip was used as the negative electrode to locally oxidize silicon and write the oxide pattern 'C-Tube'.

10. REFERENCES

Anderson, P. W., Thouless, D. J., Abrahams, E. and Fisher, D. S., 1980, New method for a scaling theory of localization, *Physical Review B*, **22**, 3519.

Avouris, Ph., Hertel, T. and Martel R., 1997, Atomic force microscope tip-induced local oxidation of silicon: kinetics, mechanism, and nanofabrication, *Applied Physics Letters*, **71(2)**, 285-287.

Bachtold et al., 1998a, Contacting carbon nanotubes selectively with low-ohmic contacts for four-probe electric measurements, *Applied Physics Letters*, **73**, 274-276.

Bachtold et al., 1998b, Electrical properties of single carbon nanotubes, *Proceedings of the XII International Winterschool on Electronic Properties of Novel Materials, "Molecular Nanostructures"*, Kirchberg, Austria.

Baumgartner, G., Carrard, M., Zuppiroli, L., Bacsa, W., de Heer, Walt A., and Forró, L., 1997, Hall effect and magnetoresistance of carbon nanotube films, Physical Review B **55** , 6704-6707.

Beenakker, C. W. J. and van Houten, H. 1991, Quantum transport in semiconductor nanostructures, Solid State Physics, vol. 44 eds H. Ehrenreich and D. Turnbull (New-York, Academic Press) 1-229.

Bergman, G., 1984, Weak localization in thin film, *Physics Reports* **107**, 1.

Bezryadin, A., Verschueren, A. R. M., Tans, S. J., Dekker, C., 1998, Multiprobe Transport Experiments on Individual Single-Wall Carbon Nanotubes, *Physical Review Letters*, **80**, 4036-4039.

Bockrath, M., Cobden, D. H., McEuen, P. L., Chopra, N. G., Zettl, A., Thess, A., and Smalley, R. E., 1997, Single-Electron Transport in Ropes of Carbon Nanotubes, *Science*, **275**, 1922-1925.

Bryson, J. W. et al., 1995, Protein design: A hierarchic approach, *Science* **270**, 935-941.

Chico, L. *et al.*, 1996, Quantum conductance of carbon nanotubes with defects, *Physical Review B* **54**, 2600-2606.

Creighton, T. E., 1992, *Protein folding* (Freeman).

Dai, H., Wong, E. W. & Lieber, C. M., 1996, Probing electrical transport in nanomaterials: conductivity of individual carbon nanotubes, *Science* **272**, 523-526.

Dai, H., Hafner, J. H., Rinzler, A. G., Colbert, D. T. and Smalley, R. E., 1996, Nanotubes as nanoprobes in scanning probe microscopy, *Nature*, **384**, 147-150.

Dai, H., Franklin, N., Han, J., 1998, Exploiting the properties of carbon nanotubes for nanolithography, *Applied Physic Letters*, **73**, 1508-1510.

Delaney, P., Choi, H. J., Ihm, J., Louie, S. G. and Cohen, M. L., 1998, Broken symmetry and pseudogaps in ropes of carbon nanotubes, *Nature*, **391**, 466-468.

See for example: Datta, S., 1995, *Electronic transport in mesoscopic systems* (Cambridge University Press, Cambridge).

Dresselhaus, M. S., Dresselhaus, G. and Eklund, P. C., 1996, *Science of fullerenes and carbon nanotubes* (Academic Press, San Diego).

Economou, E. N., 1983, *Green's function in quantum physics* (Springer-Verlag, New-York).

Ebbesen, T. W. *et al.*, 1996, Electrical conductivity of individual carbon nanotubes. *Nature* **382**, 54-56.

Fischer, J. E., Dai, H., Thess, A., Lee, R., Hanjani, N. M., Dehaas, D. L. and Smalley, R. E., 1997, Metallic resistivity in crystalline ropes of single-wall carbon nanotubes, *Physical Review B* **55**, 4921-4924.

Frank, S., Poncharal, P., Wang, Z. L., and de Heer, W. A., 1998, *Carbon nanotube quantum resistors*, *Science* **289** (1998) 1744-1746.

Hertel, T., Martel, R. and Avouris, Ph., 1998a, Manipulation of individual carbon nanotubes and their interaction with surfaces, *Journal of Physical Chemistry B* **102**, 910-915.

Hertel, T., Walkup, R. E. and Avouris, Ph., 1998b, Deformation of carbon nanotubes by surface van der Waals forces, *Physical Review B.*, 58, 13870-13873.

Journet, C. and Bernier, P., 1998, Production of carbon nanotubes, *Applied Physics A*, **67**, 1-9.

Kane, C. L. and Mele, E. J., 1997, Size, shape and low energy electronic structure of carbon nanotubes, *Physical Review Letters*, **78**, 1932-1935.

Landauer, R., 1957, *IBM Journal of Research Development*, 1, 223.

Landauer, R., 1992, Conductance from transmission: common sense points, *Physica Scripta*, **T42**, 110.

Langer, L. *et al.*, 1996, Quantum transport in a multi-walled carbon nanotube, *Physical Review Letters* **76**, 479-482.

Liu, J. et al., 1997, Fullerene 'crop circles', *Nature* **385**, 780-781.

Martel, R., Shea, H. R. and Avouris, Ph., 1998a, Rings of single-wall carbon nanotubes, to be published.

Martel, R., Schmidt, T., Shea, H. R., Hertel, T. and Avouris, Ph., 1998b, Single and multi-wall carbon nanotube field-effect transistors, *Applied Physics Letters*, **73**, 2447-2449.

Mintmire, J. W., Dunlap, B. I., White, C. T., 1992, Are fullerene tubules metallic?, *Physical Review Letters*, **68**, 631-634.

Odom, T. W., Huang, J. -L., Kim, P. and Lieber, C. M., 1998, Atomic structure and electronic properties of single-walled carbon nanotubes, *Nature* **391**, 62-64.

Rochefort, A., Salahub, D. R. and Avouris, Ph., 1998a, The effect of structural distortions on the electronic structure of carbon nanotubes, *Chemical Physics Letters*, **297**, 45-50.

Rochefort, A., Lesage, F., Salahub, D. R. and Avouris, Ph., 1998b, submitted.

Saito, R., Dresselhaus, G., and Dresselhaus, M.S., 1998, *Physical properties of carbon nanotubes* (Imperial College Press, London).

Seri, T., and Ando, T., 1997, Boltzmann conductivity of a carbon nanotube in magnetic fields, *Journal of the Physical Society, Japan,* **66**, 169-173.

Shea, H. R., Martel, R., Hertel, T. Schmidt, T and Avouris, Ph., 1998a, Manipulation of carbon nanotubes and properties of nanotube field-effect transistors and rings, Journal of Microelectronic Engineering, *in press.*

Shea, H. R. Martel, R., and Avouris, Ph., 1998b, to be published.

Suslick, K. S., Editor, 1988, *Ultrasound: Its chemical, physical and biological effects* (VCH Publishers, Weinheim).

Tans, S. J., Devoret, M. H., Dai, H., Thess, A., Smalley, R. E., Geerligs, L. J., and Dekker, C., 1997, Individual single-wall carbon nanotubes as quantum wires. *Nature*, **386**, 474-477.

Tans, S. J., Verschueren, A. R. M. and Dekker, C., 1998, Room-temperature transistor based on a single carbon nanotube, *Nature*, **393**, 49-52.

Tersoff, J. and Ruoff, R. S., 1994, Structural properties of a carbon-nanotube crystal, *Physical Review Letters* **73**, 676-679.

Thess, A. et al., 1996, Crystalline ropes of metallic nanotubes, *Science* **273**, 483-487.

Thian, W. *et al.*, 1998, Conductance spectra of molecular wires, *Journal of Chemical Physics,* **109**, 2874-2882.

Treacy, M. M., Ebbesen, T. W. and Gibson, J. M., 1996, Exceptionally high Young's modulus observed for individual carbon nanotubes. *Nature* **382**, 678-680.

Voit, J. , 1994, One-dimensional Fermi liquids, *Reports on Progress in Physics,* **57**, 977-1116.

Washburn, S. and Webb, R. A., 1986, Aharonov-Bohn effect in normal metal: quantum coherence and transport, *Advances in physics*, **35**, 375-422.

Wildöer, J. W. G., Vencma, L. C., Rinzler, A. G., Smalley, R. E. and Dekker, C., 1998, Electronic structure of atomically resolved carbon nanotubes, *Nature* **391**, 59-62.

Wong, E. W., Sheehan and P. E., Lieber, C.M., 1997, Nanobeam Mechanics: Elasticity, Strength, and Toughness of Nanorods and Nanotubes, *Science,* **277**, 1971-1975.

12 Mono-sized Carbon Nanotubes in AFI Crystal

Z. K. Tang, H. D. Sun and J. N. Wang
Department of Physics, Hong Kong University of Science and Technology, Clear Water Bay, Kowloon, Hong Kong

ABSTRACT

We report an alternative approach to the synthesis of *mono-sized* and well-aligned single wall carbon nanotubes (SWCNs). The nanotubes are accommodated in 1-nm-sized channels of porous aluminophosphate crystal by pyrolysis of tripropylamine molecules in the channels. They are characterized through transmission electron microscopy and polarized Micro-Raman scattering. Electrical transport measurements showed that the SWCN is an intrinsic semiconductor with band-gap energy of about 50 meV. The fabrication of mono-sized SWCNs in three-dimensional ordered and nano-sized channels represents an important step towards the development and applications of carbon nanotube materials.

INTRODUCTION

Carbon nanotubes were discovered from the cathode product in carbon-arc discharge method similar to that used for fullerene preparation (Iijima, 1991). This discovery triggered extensive theoretical and experimental researches, and led to a rapidly developing research field. The nanotube system offers new prospects to fundamental as well as nano-technological applications. In the earlier works, the nanotube products are mostly including fragments of graphene sheets, carbon clusters and amorphous carbons. Recently, purification process to obtain nanotubes out of the mixture carbon specimens has been gradually established. An important recent advance in carbon nanotube science is the synthesis of single-walled carbon nanotubes (SWCNs) in high yield using the laser ablation (Guo, 1995) and electric arc technique (Journet, 1997). In each case a small amount of transition metal is added to the carbon target as a catalyst. The ability of SWCN synthesis brings the experimental situation much closer to that of the theoretical models. However, it remains to be a challenge to produce aligned and *mono-sized* SWCNs with well-defined chirality.

From theoretical point of view, a SWCN is interesting as the embodiment of a one-dimensional periodic structure along the tube axis. Depending on the rolling direction of the graphene sheet, a carbon nanotube has three possible structures: zigzag, armchair and chiral, which can uniquely be represented by a so-called roll-up vector (n, m) (Dresselhaus, 1995). The electronic property of a carbon nanotube is expected to be purely dependent on its geometrical structure (chirality and tube size). An armchair nanotube (n, n) is metal, while a zigzag nanotube $(n, 0)$ is a semiconductor with a narrow band gap (when n is the multiply of three) or with a moderate band-gap (for other n) (Blasé, 1994; Mintmire, 1995). Independent of whether the nanotubes are conducting or semiconducting, the energy bands are expected to show one-dimensional characteristic density-of-state with a $(E - E_i)^{-1/2}$ type singularity at its band maximum energy E_i (Dresselhaus, 1996). However, experimental measurements to test these remarkable expectations of the electronic structure are not easy because of the difficulty in fabricating *mono-sized* carbon nanotubes with well-defined structural symmetry. Another technical challenge is to make a good electric electrode to such a small tube. Ups to now, the experimental studies have been mainly limited to nanotube bundles (Thess, 1996; Lee, 1997) and individual multi-walled carbon nanotubes (Ebbesen, 1996; Langer, 1996), which only qualitatively show semiconductor or metal properties of the nanotubes. Recently, measurements using two-probe configuration showed that individual SWCNs behave as molecular wires, exhibiting quantum effects at low temperature (Tans, 1997; Bockrath, 1997). For more controlled characterization of these novel electronic properties, separated, aligned and mono-sized nanotubes with a well-defined chirality are essential.

In this paper, we report an alternative approach to the synthesis of *mono-sized* and parallel-aligned SWCNs in the 1 nm-sized channels of zeolite AlPO$_4$-5 (AFI) crystals, using pyrolysis of tripropylamine molecules. As the nanotubes are well aligned in the macro-sized AFI crystal, we can measure polarised Raman spectra and directly measure the conductivity for the nanotube sample, which provide us a lot of information on structural symmetry and electronic states.

EXPERIMENTAL

AFI is a kind of microporous aluminophosphate crystal. The framework of the AFI crystal is constructed of alternative tetrahedra of $(AlO_4)^-$ and $(PO_4)^+$, which form parallel open channels, packed in hexagonal structure. The co-ordinate diameter of the channels is 10.1 Å (inner diameter 7.3 Å), and the separation distance between two neighbouring channels is 13.7 Å. In this research, the AFI crystal with elongated hexagonal prisms of dimensions 110 μm in cross-section diameter and 300 μm in length were used as the hosts to encapsulate the SWCNs. Hydrocarbon molecules of tripropylamine (TPA) were used as the starting material to produce the carbon nanotubes in the AFI channels. The TPA molecules were introduced into the channels during the AFI crystal growth, and they were pyrolysed in a vacuum of 10^{-4} Torr at temperature 350 ~ 450 °C. Carbon nanotubes were formed in the AFI channels when the pyrolysed carbons were thermally treated at 500 ~ 800 °C in vacuum. The AFI single crystals were carefully selected, as the formation of the SWCNs is very sensitive to the quality of the AFI crystal. A good sample of the carbon-nanotube contained AFI crystal (C-AFI) behaves as a good polariser with high absorption for the light polarized parallel to the channel direction ($E//c$) and with a high transparency for $E \perp c$, in consisting with the one-dimensional characteristics of the nanotubes.

RESULTS AND DISCUSSIONS

A. Transmission-Electron-Microscopy Observation

Transmission-Electron-Microscopy (TEM) is a powerful tool to observe the structure of freestanding carbon nanotubes. However, it is hard to directly observe the nanotubes accommodated in the AFI channels because of the facts: (a) The AFI framework is unstable under electron beam illumination, and (b), in comparison with the heavy elements (Al and P) of the AFI host, carbon element is too light to give a reasonable contrast in TEM image. In our experiment, we prepared the samples for TEM observations by removing the AFI host framework

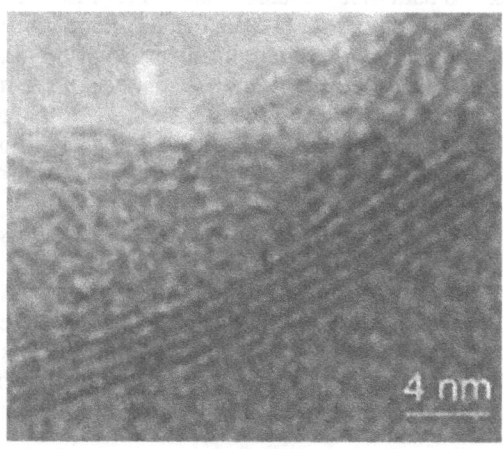

Fig. 1 Transmission electron microscope (TEM) image shows the SWCNs which is slightly bent. The TEM image was taken after the AFI framework being removed using HCl acid, and the SWCNs being exposed in free space.

using acid etch, and depositing the separated nanotube phase onto a carbon mesh. Figure 1 shows the TEM image of the carbon nanotubes. The image was taken using a Philips CM200 at 200 kV. Smoothly bent wire-like carbon specimens, with separation distance of about 7 Å, are seen in the image. We attribute the wire-like carbon specimens to carbon nanotubes, which were bundled together after removing the AFI framework. Because the small size of the AFI channel, geometrically only single-walled carbon nanotubes can be formed in the channels. Considering possible physical interaction between the nanotubes and the channel walls, the diameter of the SWCN is expected to be less than 6 Å, which is smaller than reported SWCNs produced using other methods. The structure of the small tubes is stable when they are confined in the channels of AFI crystal, but they become unstable when they are exposed outside the channels, especially under electron-beam irradiation. This fact indicates that the one-dimensional channels of AFI crystal play an important role in stabilising small carbon nanotubes.

B. Polarized Raman spectra

Since nanotubes are well aligned in the crystal AFI channels. It makes identify the symmetry of the carbon nanotubes by measuring polarized Raman spectra. Curves (a) and (b) in Fig. 2 show the polarized Raman spectra of the SWCNs measured using 632.8-nm line of a He-Ne laser as excitation for the polarization configurations of ZZ and XZ, respectively, where Z and X denote the polarization directions along and perpendicular to the tube direction. The first (second) notation is for the incident laser (scattering light). As seen in the figure, the

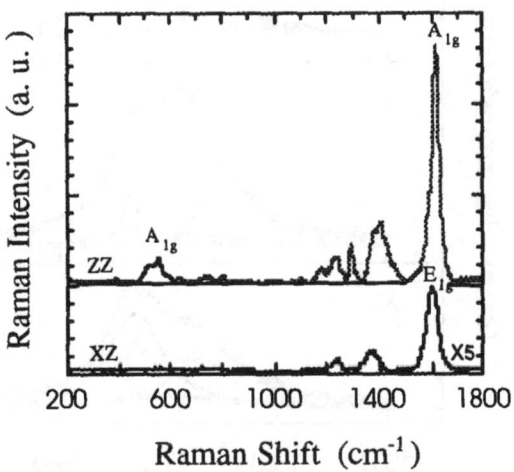

Raman Shift (cm^{-1})

FIG. 2 Polarized Raman spectra of single-wall carbon nanotubes formed in the channels of an AFI crystal in the three configurations ZZ and XZ. The long axis of the AFI crystal is along the Z axis

Raman intensity measured in the ZZ configuration is much stronger than that in XZ configuration, indicating that the carbon specimen in the AFI channels are of one-dimensional characteristics. In the ZZ configuration, a strong Raman signal at 1615 cm^{-1}, with weak shoulders at low frequency side, dominates the spectrum. These Raman peak and shoulders have been attributed to the graphite-like longitudinal and transverse optical phonon modes branched off from the Γ point to higher and lower energies with discrete wave vectors (Kasuya, 1997). In low frequency region (400 ~ 800 cm^{-1}), there is a significant Raman signals centred at 530 cm^{-1}. Since the frequency is in the silent region for graphite and other carbon materials, this Raman signal can be assigned to the radial-breathing mode, which is special to the nanotube geometry. The breathing mode is not dependent on the chirality of nanotubes, but sensitive to the nanotube diameter (Jishi, 1993; Saito,

1998). The frequency of the mode is approximately a linear function of the reciprocal of the nanotube diameter. From the well-established data for (10, 10) armchair nanotubes (Jishi, 1993; Saito, 1998), the frequency of the breathing mode can be estimated to be 448 cm^{-1} for the SWCNs in the AFI channels. The estimated Raman frequency is lower than the measured frequency of 530 cm^{-1}

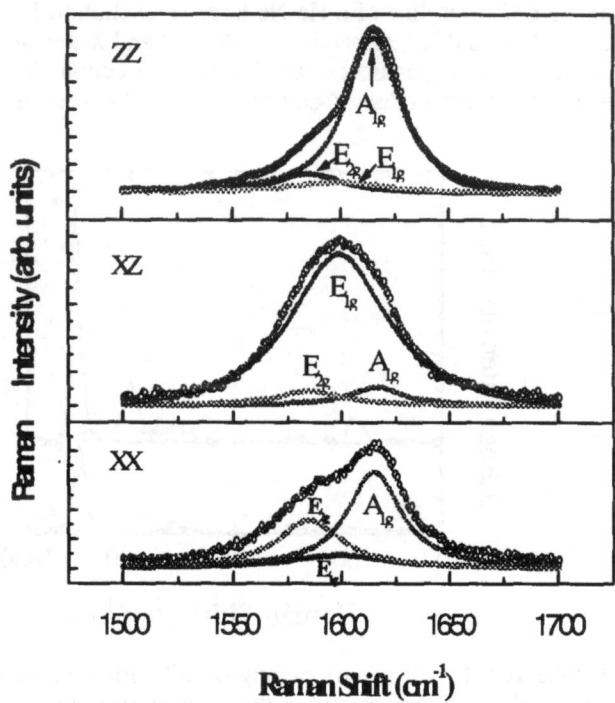

shown in Fig. 2. The up-shift of frequency may result from the physical interaction between the nanotube wall and the rigid wall of channels of AFI crystals.

Fig.3. Polarised Raman spectra near ~1600 cm^{-1}. The open circles are experimental data; the dotted curve is fitted using the Lorentzian lineshapes. The solid curves are the total intensities of three Lorentzian lineshapes.

As shown in Fig. 2, the relative intensities of the Raman signals is sensitive to the polarization configurations. The assignment of the Raman modes in the intermediate frequency region (1000 ~ 1500 cm^{-1}) is still ambiguous. In the following we focus our discussion of the polarization properties to the breathing mode around 530 cm^{-1} and the graphite-like modes around 1600 cm^{-1}. As seen in Fig. 2, the breathing mode can be observed in the ZZ polarization configuration but disappeared in the depolarized spectrum of *XZ* configuration. On the other hand, the Raman signal of the graphite-like modes is peaked at 1615 cm⁻¹ with a shoulder at 1585 cm⁻¹ in the *ZZ* configuration. This strong 1615 cm⁻¹ line is replaced by a new line located at 1599 cm⁻¹ in the *XZ* configuration. Fig. 3 shows more detailed polarisation dependence for the Raman lines near 1600 cm⁻¹ in three different configurations. Raman bands measured in these three configurations are fitted using the Lorentzian lineshape, where the open circles are experimental data, the dotted curves are fitting Lorenzian shapes and the solid curves are the total intensities of the Lorenzian lines, respectively. As seen from the figure, the Raman band in this frequency region is contributed from three lines named P_1, P_2 and P_3, respectively. In the *ZZ* configuration, P_1 is the main line while P_3 appears only as a weak shoulder and the contribution of P_2 line is negligible. In *XZ* polarization configuration, the P_2 line becomes the main peak and the magnitudes of the P_1 and P_3 lines contributing to the Raman band are very small. In *XX* polarization configuration, the P_2 line has only a negligible small intensity, P_1 and P_3 are dominated in the Raman band.

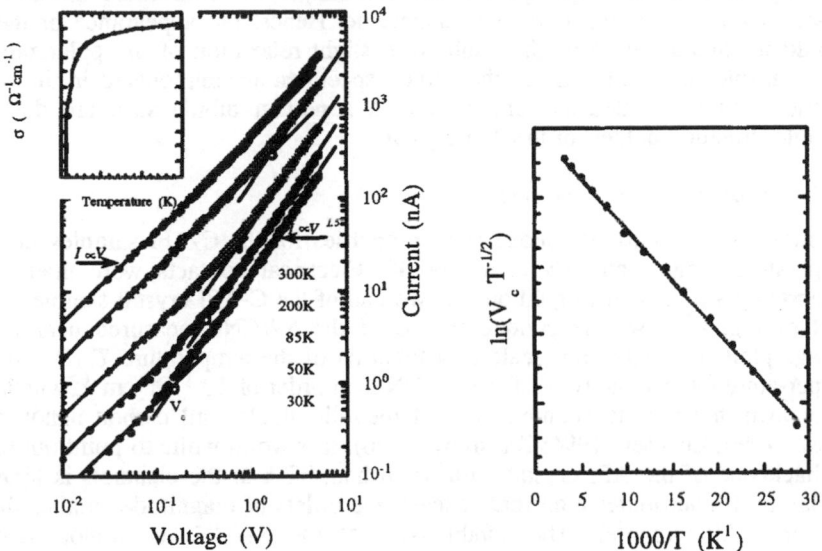

Fig.4. (Left) *I-V* curves measured at different temperatures (open circles) plotted in a log-log scale. The solid straight lines are the fits with $I \propto V$ in low bias region and $I \propto V^{3/2}$ in higher bias region. The inset shows the temperature dependence of the conductivity measured near zero bias.

Fig.5. (Right) Temperature dependence of the cross-over voltage, V_c (solid circles), where $\ln(V_cT^{1/2})$ is plotted as a function of $1/T$. The solid line is the corresponding linear fit.

Among the three structure symmetries of carbon nanotubes, the armchair structure (n, n) can be excluded from our C-AFI sample, as the carbon nanotube in the AFI channel was observed to be a narrow-gap semiconductor (Tang, 1998). The group theory for carbon nanotubes predicts that there are 15 Raman-active vibration modes at $k = 0$ for all zigzag $(n, 0)$ tubules, containing the symmetries of $3A_{1g}$, $6E_{1g}$ and $6E_{2g}$ (Dresselhaus, 1995; Eklund, 1995; Jishi, 1993). According to the polarization selection rule of these modes, A_{1g} mode is allowed in the ZZ and the XX configurations, but it is forbidden in the XZ configuration; E_{1g} mode is allowed only in the XZ configuration while the E_{2g} mode is allowed only in the XX configuration. Comparing the polarization selection rules of these modes to the experimental data showed in Fig. 2 and Fig. 3, we can reasonably assign the low-frequency Raman line to the A_{1g} breathing mode, and the bands of P_1, P_2 and P_3 located in the high frequency region to the A_{1g}, E_{1g} and E_{2g}, respectively. It is worth to point out, however, that the E_{2g} mode is also present in the ZZ and XZ configurations and A_{1g} in the XZ configuration, as weak shoulders. These modes should be absent according to their polarisation selection rules. The relative intensities of these symmetry-forbidden modes in Raman spectra were observed to strongly depend on sample quality. In a bad sample, these forbidden modes can be seen with even stronger relative magnitude. Hence, the appearance of these forbidden Raman modes might result from slight relaxation of the polarization selection rule because not all of the carbon specimen accommodated in the AFI channels can be treated as nanotubes with a perfect tubule structure due to possible structural defects of the AFI crystal.

C. *Electrical transport properties*

We measured the current-voltage $(I\text{-}V)$ properties for the C-AFI samples in the temperature range from 0.3 K to 300 K. Electrical contacts were made by evaporating a thin layer of gold on the two end of the C-AFI crystal sample. The inset of Fig. 4 shows the conductivity σ of the SWCNs measured near zero voltage plotted in logarithm scale as a function of the temperature T. At room temper-ature the conductivity of the SWCN is in order of 10^{-1} $\Omega^{-1}\text{cm}^{-1}$, which is lower than that reported conductivity of metallic single-wall carbon nanotubes (Thess, 1996; Ebbesen, 1996; Kasumov, 1996). It is worth while to point out that the backbone of the AFI crystal with or without TPA in the channels is highly insulating. It contributes a current at least five orders of magnitude smaller than that of C-AFI samples. The conductivity of the SWCNs is monotonically decreased with decreasing temperatures, indicating that the SWCN is of semiconductor characteristics. The typical dc $I\text{-}V$ curve (open circles) measured at different temperatures is plotted in log-log scale in Fig. 4. The solid lines are fittings with $I \propto V$ and $I \propto V^{3/2}$ as indicated in the figure. The current starts out as a linear function of the voltage, but switches to a $V^{3/2}$ behavior when the bias voltage is high or when the temperature is low. Since the SWCNs in our experiment are not intentionally doped, the conducting carriers are expected to be due to the thermally generated free carriers. Because the electrode metal and the SWCN have different work functions, a Schottky barrier can be formed at the lead-sample contact. When a bias voltage is applied, the barrier prevents one type of the thermally excited carriers (say electrons) to pass through, but allows the other type of carriers (say holes) to pass through freely. Hence, electrical conduction in the SWCN is expected to be a *single-carrier* process. A quantitative description of the experimental *I-V* curves can then be obtained by evaluating the current across the sample by using the transport equation [28,29], $J = q\mu(n+n_0)E$, and the Poisson equation $(1/r)(d(r^2E)/dr = qn/\varepsilon$, here q is the carrier charge, μ the carrier drift mobility, E the applied electric field, ε the static electric constant, n and n_0 are the free carrier density injected from the electrode and the thermally generated carrier density, respectively. At high temperature or low bias voltage, $n_0 \gg n$, the current is simply given by the transport equation as $I_\Omega = I_{\Omega 0}\,\mu(n + n_0)\,V$, which is ohm current. At low temperature or under high bias, n is comparable with n_0. Solving the transport equation and the Poisson equation, we get $I_{SCL} = I_{s0}\,\mu n_0^{1/2} V^{3/2}$, which is so-called space-charge-limited (SCL) current. At a given temperature, the total current of the sample is due to the superposition of the I_Ω and the I_{SCL}. We expect the current to be dominated by I_Ω at low bias and high temperature and by I_{SCL} at high bias and low temperature. The solid lines in Fig. 4 show the calculated currents at different temperatures. The good agreement between experimental data and calculation also indicates that the SWCN confined in the AFI channel is an intrinsic semiconductor. The crossover voltage V_c, defined as the voltage at which $I_\Omega = I_{SCL}$, is simply given by $V_c = (I_{\Omega 0}/I_{s0})^2 n_0$, where $I_{\Omega 0}/I_{s0}$ is a constant for a given electrode contact configuration. It is interesting to notice that the V_c is linearly proportional to the thermal carrier density n_0, but is independent on the carrier mobility μ. The simple dependence of V_c on the carrier density n_0 offers us a ground to determine the band-gap for the SWCNs. As seen in Fig. 4, the V_c shifts towards higher value as increasing temperature, because of the increase of n_0 with the temperature. The V_c obtained by fitting the *I-V* curves is plotted in Fig.5 (open circles) as a function of the temperature. Using one-dimensional characteristic density-of-state $(E-E_i)^{-1/2}$, the temperature dependence of the thermal-generated carrier density n_0 can be calculated as $n_0 = AT^{1/2} \exp(-E_g/2k_B T)$, here A is a constant, E_g the band-gap energy, k_B the Boltzman constant and T the temperature. The solid curve in Fig. 5 is the plot of $\ln(n_0/T^{1/2})$ against $1/T$, which gives a linear line. The band-gap energy E_g can then be determined from the line slope. This relationship is shown in Fig. 5. The experimental data V_c (open circles) is fitted well using an energy gap of 52 meV.

In summary, we have fabricated *mono-sized* SWCNs with well-defined structure symmetry. They were characterised through TEM, Raman scattering and electrical transport measurements. Our results would open a door to further detailed studies on the intrinsic properties of semiconducting carbon nanotubes, presently under way.

Acknowledgement   We are grateful to M. M. T. Loy, P. Sheng and G. K. L. Wong for their encouragement and valuable comments. In particular, H.D.Sun is grateful to Prof. G. K. L. Wong for introducing him to this research area. The authors also thank Dr. C. T. Chen for his theoretical advice and useful discussion. The TEM image was taken by Dr. N. Wang. This research was supported by the RGC Committee of Hong Kong, and the EHIA program from HKUST.

REFERENCES

Blasé, X., et al., 1994, Physics. Review. Letters, 72, pp. 1878-1881.
Bockrath, M., et al., 1997, Science, 275, pp. 1922-1925.
Dresselhaus, M. S., D. Dresselhaus, D., and R. Saito, R., 1995, Carbon, 33, pp. 883-91.
Dresselhaus, M. S., Dresselhaus, D., and Eklund,P. C., 1996, in Science of Fullerens and Carbon Nanotubes, (Academic, New York).
Ebbesen, T. W., et al., 1996, Nature, 382, pp. 54-56.
Eklund, P. C., Holden, J. M., and Jishi, R. A., 1995, Carbon, 33, pp. 959-972.
Guo, T., et al., 1995, Chemical Physics Letters, 243, pp. 49-54.
Iijima, S., 1991, Nature, 354, pp. 56-58.
Jishi, R. A., et al., 1993, Chemical Physics Letters, 209, pp. 77-82.
Journet, C., et al., 1997, Nature, 388, pp. 756-760.
Kasumov, A. Y., et al., 1996, Europhysics Letters, 34, pp. 429-432.
Kasuya, A., et at., 1997, Physics Review Letters, 78, 4434-4437.
Langer, L., et al., 1996, Physics Review Letters, 76, pp. 479-482.
Lee, R. S., et al., 1997, Nature, 388, pp. 255-257.
Mintmire, J. W., and White, C. T., 1995, Carbon, 33, pp. 893-902.
Saito, R., et al., 1998, Physics Review, B 57, pp. 4145-4153.
Tang, Z. K., et al., 1998, Applied Physics Letters, 73, pp. 2287-2289
Tans, S. J., et al., 1997, Nature, 386, pp. 474-477.
Thess, A., et al., 1996, Science, 273, pp. 483-487.

13 Preparation and Properties of Processible Polyacetylene-wrapped Carbon Nanotubes

Ben Zhong Tang and Hongyao Xu
Department of Chemistry, The Hong Kong University of Science and Technology, Clear Water Bay, Kowloon, Hong Kong

ABSTRACT

Carbon nanotubes are intractable, which hampers the progress in the nanotube research and limits the scope of their practical applications. In this work, we developed a simple process for solubilizing the carbon nanotubes and discovered novel electronic, optical, and mechanical properties of the nanomaterials. The nanotube-containing polyphenylacetylenes (NT–PPAs) are prepared by *in situ* polymerisations of phenylacetylene catalysed by WCl_6–Ph_4Sn and $[Rh(nbd)Cl]_2$ in the presence of the nanotubes. The NT–PPAs are characterised by GPC, NMR, UV, FL, TGA, SEM, TEM, and XRD, and it is found that the nanotubes in the NT–PPAs are helically wrapped by the PPA chains. The short nanotubes thickly wrapped in the PPA chains are soluble in common organic solvents including THF, toluene, chloroform, and 1,4-dioxane. The NT–PPAs are macroscopically processible, and shearing the concentrated NT–PPA solutions readily aligns the nanotubes along the direction of the applied mechanical force. The fluorescence of PPA is largely quenched by the nanotubes, possibly *via* photoinduced charge transfer process from the PPA chains to the nanotube shells. The nanotubes dramatically stabilise the PPA chains against harsh laser irradiation, and the NT–PPA solutions effectively limit intense optical pulses, with the saturation fluence easily tunable by varying the nanotube contents.

1 INTRODUCTION

The discovery of carbon nanotubes and the prospect of developing carbon-based nanomaterials excited worldwide interest among researchers (Ebbesen, 1997; Endo *et al.*, 1996). Thanks to the enthusiastic research efforts of scientists, the nanotubes have quickly proven to possess exotic materials properties such as high mechanical strength and flexibility (Falvo *et al.*, 1997; Treacy *et al.*, 1996) and diameter- and chirality-dependent electrical conductivity (Odom *et al.*, 1998; Wildoer *et al.*, 1998). The intractability of the nanotubes, however, poses an obstacle to further development in nanotube science and significantly limits the scope of their practical applications. No suitable solvents for the nanotubes have been found, and because of the wet nature of chemical research, this has largely blocked chemists' entry to the area of nanotube research. It is envisioned that solubilization of the carbon nanotubes will not only endow the nanotubes with processibility but also open up new avenues in the nanotube research.

The nanotubes are sometimes referred to as "quantum wires" (Ebbesen, 1997) and polyacetylenes are the best-known "synthetic metals" (Taliani *et al.*, 1993). Compounding the nanotubes with soluble polyacetylenes is of interest because the resulting nanocomposites may possess novel electronic, optical, magnetic, and mechanical properties. Furthermore, the ability to orient the polyacetylene chains by external forces (Akagi and Shirakawa, 1996; Kong and Tang, 1998; Tang *et al.*, 1998) may allow alignment of the nanotubes in the polyacetylene nanocomposites by macroscopic means such as mechanical perturbations. In this work, we chose polyphenylacetylene (PPA), a soluble photoconductive polyacetylene (Kang *et al.*, 1984) as a model polymer. We describe here a simple method for the preparation of soluble nanotube-containing PPAs (NT–PPAs). We demonstrate that the nanotubes in the NT–PPA solutions can be easily aligned by mechanical shear and that the NT–PPA solutions effectively limit intense 532-nm laser pulses.

2 EXPERIMENTAL SECTION

In a typical run of the polymerisation reactions catalysed by WCl_6 and Ph_4Sn, into a baked 20-mL Schlenk tube under an atmosphere of dry nitrogen were added 100 mg of well-ground multiwalled carbon nanotubes (purchased from MER and further ground in our laboratory), 95 mg of WCl_6, 102 mg of Ph_4Sn, 5 mL of toluene, and 0.55 mL of phenylacetylene (PA; all from Aldrich). The mixture was stirred at room temperature for 24 hours. Toluene (5 mL) was then added to the Schlenk tube and the diluted reaction mixture was filtered by a cotton filter to remove large, insoluble nanotube particles. The soluble filtrate was added dropwise into 250 mL of methanol under stirring to precipitate the polymeric product. After standing overnight, the precipitant was isolated using a Gooch-type crucible filter and dried in a vacuum oven. The product was redissolved in tetrahydrofuran (THF) and the resulting solution was centrifuged at 2 000 rpm for 16 minutes. The homogeneous supernatant was added through a cotton filter into hexane under stirring. The precipitant was collected and then dried in *vacuo* at 40 °C to a constant weight. The NT–PPA was obtained in an isolated yield of 45.3 wt %.

The polymerisation reactions were also catalysed by $[Rh(nbd)Cl]_2$ (nbd = 2,5-norbornadiene). The large amount of insoluble nanotube particles recovered from the polymerisation reaction using a high feed ratio of nanotube/PA (102 mg/0.9 M) (*cf.*, Table 1, no. 3) were repeatedly washed by THF. The solvent was tinged yellowish in the beginning of the washing but became colourless at the end. The particles were further soaked in THF for a period of 20 days but the solvent remained uncoloured. The particles separated from the supernatant were washed with fresh THF again and then dried under vacuum to a constant weight (108 mg).

3 RESULTS AND DISCUSSION

3.1 Synthesis and Characterisation

During our research program on the development of fullerene-based optical materials (Peng *et al.*, 1997; Tang, 1996; Tang *et al.*, 1997; Tang and Peng, 1997), we have found that C_{60} can copolymerise with alkynes such as PA (Tang, 1998; Xu and Tang, 1998). Noticing that the nanotubes are conceptually the cylinders of curled graphene sheets capped by the fullerene hemispheres, we conducted a polymerisation of PA catalysed by WCl_6–Ph_4Sn in the presence of the nanotubes.

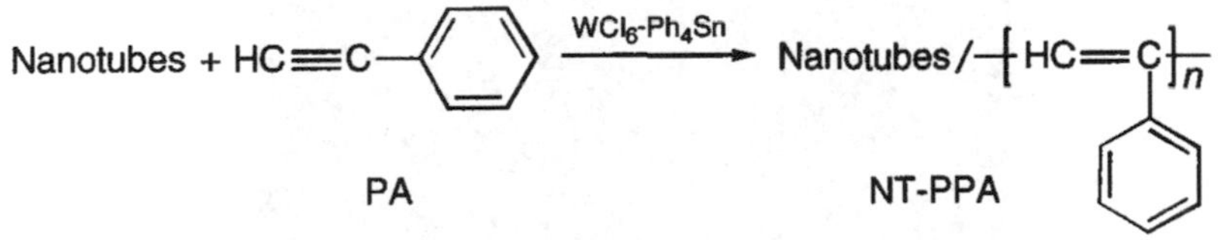

Scheme 1 **Polymerisation of phenylacetylene in the presence of carbon nanotubes.**

The polymer product isolated from the W-catalysed polymerisation reaction was characterised by scanning electron microscope (SEM), which shows numerous short and somewhat bent tubules with lengths of a few hundred nanometers (Figure 1A). Such tubular images are not observed in the SEM micrograph of the pure PPA obtained from the control experiment, suggesting that the nanotubes are compounded with PPA *in situ* during the polymerisation reaction (Scheme 1). It is not surprising that all the tubules shown in Figure 1 are short because the long nanotubes have been effectively removed by the repeated filtration during the process of polymer purification. The tubules are bent because of the mechanical force exerted by the polymer matrix, indicative of the high flexibility of the nanotubes (Falvo *et al.*, 1997; Treacy *et al.*, 1996). At high magnification, some tubules are found to be linked up, probably by the PPA chains. The thin string marked by letter "P" in Figure 1B looks like a stretched PPA "rope" with its two ends attaching to two interconnected nanotubes. Closer inspection of the micrograph, especially in those regions marked by the arrows, reveals that the tubules are surrounded by faint veils, intimating that the

124   *Tang and Xu*

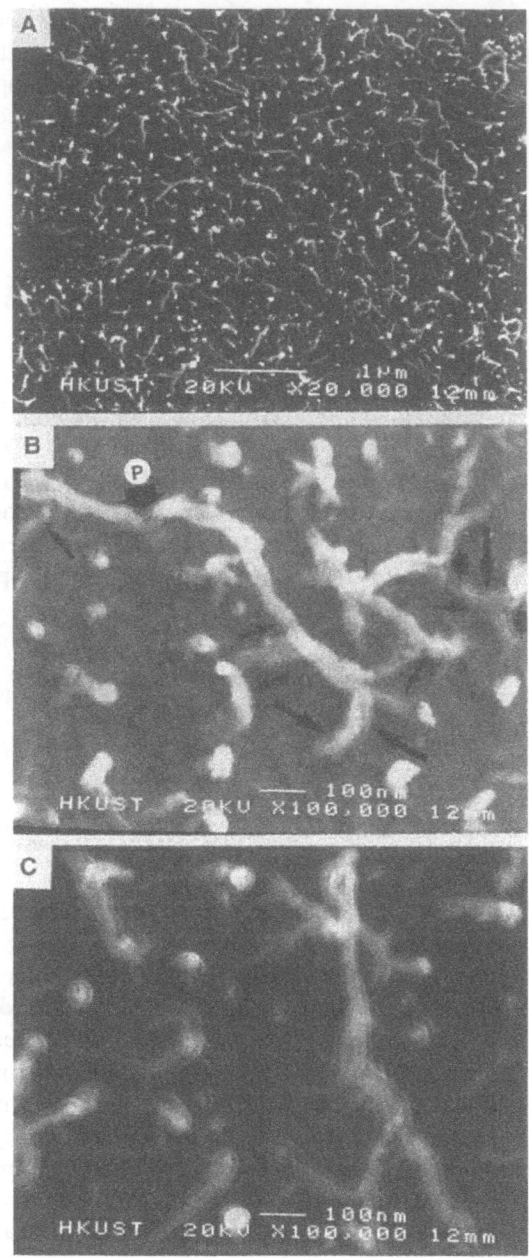

Figure 1 Carbon nanotubes dispersed in the polyphenylacetylene matrix (sample from Table 1, no. 1); micrographs taken on a JEOL 6300L ultra-high resolution scanning electron microscope operating at an accelerating voltage of 20 kV.

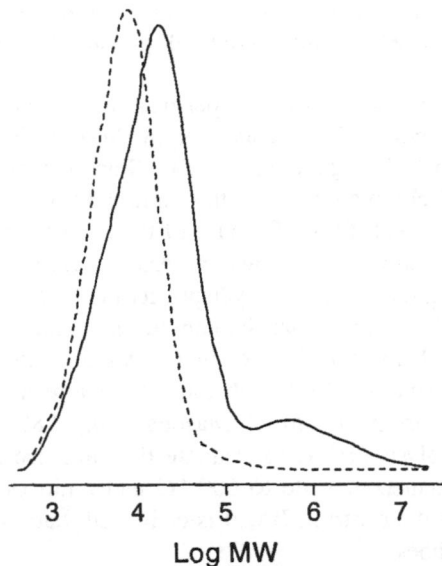

Log MW

Figure 2 Gel permeation chromatograms of NT–PPA (solid line; sample from Table 1, no. 1) and PPA (dotted line) recorded on a Waters 510 gel permeation chromatography system with a set of Styragel columns (HT3, HT4, HT6) covering an MW range of 10^2–10^7 g/mol.

Table 1 Preparation of NT–PPAs; the polymerisation reactions were carried out under an atmosphere of dry nitrogen at room temperature in toluene (5 mL) for 24 hours with a PA concentration of 0.9 mol/L, and the polymer products were purified by repeated precipitation of their toluene and THF solutions into methanol and hexane.

No.	Carbon nanotube (mg)	catalyst[a]	Polymer yield (wt %)	$M_w/10^{3b}$ (g/mol)	PDI[b]	Nanotube content[c] (wt %)
1	100	WCl$_6$–Ph$_4$Sn	45.3	172	27	6.0
2	28	WCl$_6$–Ph$_4$Sn	66.8	93	11	1.9
3	102	[Rh(nbd)Cl]$_2$	3.4			
4	20	[Rh(nbd)Cl]$_2$	50.2	137	13	2.2

[a] [WCl$_6$]$_0$ = [Ph$_4$Sn]$_0$ = 43 mM; {[Rh(nbd)Cl]$_2$}$_0$ = 2 mM. [b] Estimated by GPC calibrated with 12 monodisperse polystyrene standards (Waters); PDI: polydispersity index. [c] Estimated by TGA (Perkin-Elmer TGA 7) under nitrogen.

nanotubes are bundled or wrapped up by the PPA chains. At some spots, tree-like structures are imaged (Figure 1C), in which the "trunks" and "branches" of the isolated and interconnected nanotubes are in all probability sheathed by the "bark" of the PPA chains.

The NT–PPA shows a bimodal gel permeation chromatogram (GPC), with one large peak in the "normal" molecular weight (MW) region and another small peak in the very high MW region (Figure 2). Since the high MW peak is not observed in the GPC chromatogram of the control PPA, it should be from the nanotubes compounded with PPA. Because of the contribution of the nanotubes, the polystyrene-calibrated weight-average molecular weight (M_w) of the NT–PPA is very high (172 000 g/mol) and its polydispersity index (PDI) is extremely broad (27; Table 1, no. 1). It is known that the nanotubes are thermally very stable and do not lose any weight below 680 °C even in air (Ajayan *et al.*, 1993; Tsang *et al.*, 1993). On the other hand, the PPA, stable at room temperature notwithstanding, rapidly loses its weight from *ca.* 200 °C (Masuda *et al.*, 1985). We thus employed thermogravimetric analysis (TGA) to evaluate the nanotube content of the NT–PPA by heating the nanocomposite to 600 °C under nitrogen. By comparison with the TGA data of the control PPA, it is estimated that the NT–PPA contains 6.0 wt % of the nanotubes.

The parent carbon nanotubes are completely insoluble in THF, even with the aid of prolonged sonication. The NT–PPA, however, readily dissolves in THF, giving macroscopically homogeneous and visually transparent solutions (Figure 3), although its maximum solubility is only about two-thirds of that of the parent PPA in the same solvent. A PPA solution with a concentration of 4.00 mg/mL is yellow in colour, while a NT–PPA solution with a lower concentration (3.18 mg/mL) is orange-coloured. The deepening in colour is attributable to the dissolved nanotubes, which possess extensively conjugated π-electrons. The NT–PPA is also soluble in other common solvents including toluene, chloroform, and 1,4-dioxane, as summarised in Table 2.

Table 2 Solubility of NT–PPAs; the numbers correspond to those in Table 1.[a]

No.	THF	Toluene	CHCl$_3$	1,4-Dioxane
1	√	√	√	√
2	√	√	√	√
3		(Not determined)		
4	√	√	√	×

[a] Code: √, completely soluble; ×, partially soluble.

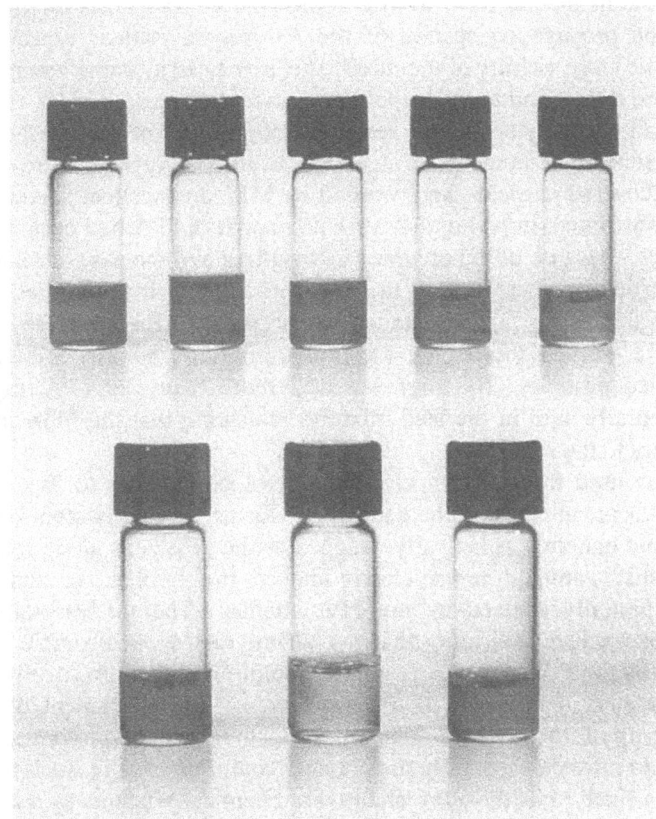

Figure 3 Homogeneous and transparent solutions of NT–PPA (sample from Table 1, no. 1) in THF with concentrations (*c*) of (upper panel from left) 19.07, 10.59, 7.15, 5.45, and 3.18 mg/mL. For comparison, THF solutions of NT–PPA and PPA with similar concentrations are given in the lower panel; *c* (mg/mL; from left): 5.45 (NT–PPA), 4.00 (PPA), 3.18 (NT–PPA).

When the amount of the nanotubes in the mixture of the polymerisation reaction decreases from 100 mg to 28 mg, the yield of the polymer product increases from 45.3% to 66.8% (Table 1, no. 2). In the absence of the nanotubes, the polymer yield is as high as 92.0%. It is thus clear that the presence of the nanotubes in the polymerisation mixture decreases the polymer yield. A similar but more pronounced "nanotube effect" is observed when [Rh(nbd)Cl]$_2$ is used as the polymerisation catalyst. The polymerisation of PA in the presence of a large amount (102 mg) of the nanotubes yields little polymer (3 4%; Table 1, no. 3). Decreasing the feed amount of the nanotubes to 20 mg dramatically increases the polymer yield to 50.2%, and the polymerisation in the absence of the nanotubes produces PPA in quantitative yield (100%). It is possible that the transition metals and the nanotubes have formed some kind of complexes of low catalytic activity. The propagation species of the PA polymerisation, especially when growing in the close vicinity of the tubule shells, may form stable complexes with the nanotubes, thus terminating the polymerisation reactions.

From all the polymerisation reactions carried out in the presence of the carbon nanotubes, fair amounts of insoluble nanotube particles were recovered. When the recovered particles were washed by THF, an excellent solvent of PPA, the solvent was tinted, implying that a small amount of PPA had been attached to the nanotubes. Because of the accuracy in weighing and the ease of handling, we chose the particles recovered from the polymerisation with the highest nanotube feed ratio (Table 1, no. 3) for detailed investigation. The recovered nanotube particles were washed thoroughly for prolonged time, but the amount of the finally isolated particles (108 mg) was still more than that (102 mg) of the nanotubes initially used in the feed mixture, indicating that the PPA chains had stuck indelibly to the nanotubes.

We thus used transmission electron microscope (TEM) to "see" how the polymer chains are attached to the nanotubes. As can be clearly seen from Figure 4A, the thinner nanotube is helically wrapped by the PPA coils along the latitudes of tubule shells. Although not so clearly imaged, the tip of the carbon nanotube seems also partially covered by the PPA chains. The thicker neighbouring nanotube looks naked, indicating that not all the recovered insoluble nanotubes are wrapped by the PPA chains. Similarly, the thinner nanotube in Figure 4B is coiled by the PPA helices. In this case, however, the attachment of the PPA chains to the tip of the nanotube is clearly imaged. Since the prolonged washing by THF should have cleaned away the polymer chains physically stuck to the tube *tip*, it is thus likely that the PPA chains are chemically bound to the fullerene hemisphere, probably by the polymerisation of PA with the C_{60} moiety (Tang, 1998; Xu and Tang, 1998). At other spots, some tubules are completely wrapped by the PPA chains, with the helices running perpendicular to the long axes of the nanotubes (Figure 4C).

To further confirm the identity of the TEM images, we carried out powder X-ray diffraction (XRD) analysis of the recovered insoluble nanotube particles. The XRD pattern of the parent carbon nanotubes exhibits an intense (002) Bragg reflection at $2\theta = 26.1°$ (Figure 5A), corresponding to the intershell spacing ($d = 3.4$ Å) of the concentric cylinders of graphitic carbon. The recovered nanotube particles show, in addition to the nanotube reflection, two new weak and broad peaks at lower angles, which, by comparison with the diffractogram of pure PPA shown in Figure 5C, are clearly the reflection peaks of the PPA chains. The XRD data thus support the correlation of the TEM images; that is, the coils wrapping around the tubule shells and attaching to the tubule tips are the amorphous PPA chains.

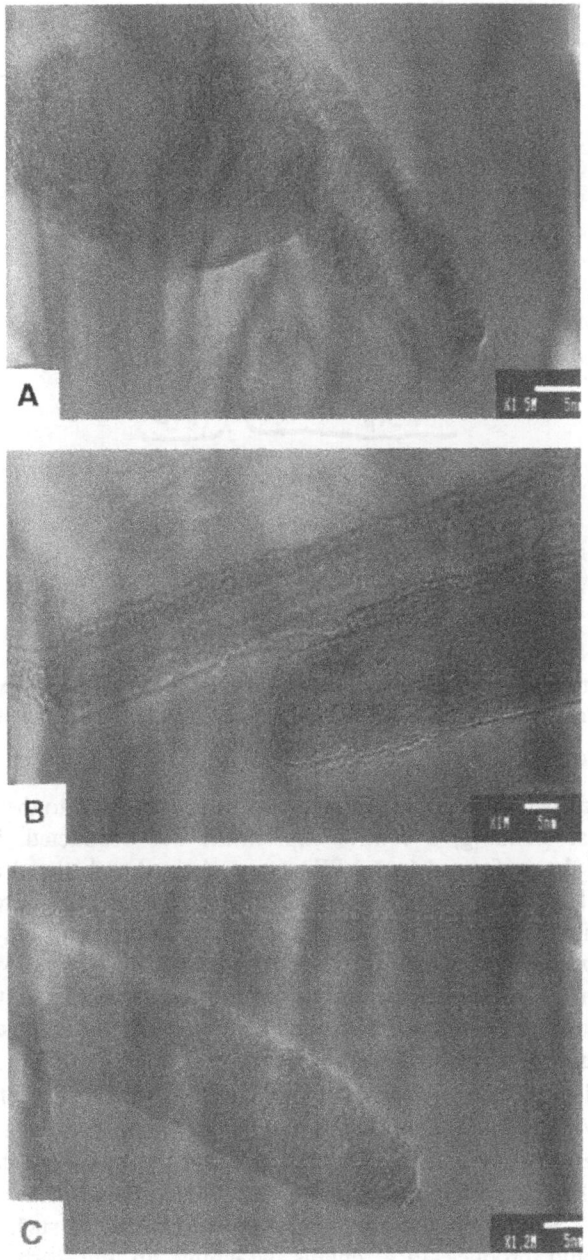

Figure 4 Wrapping of PPA chains around carbon nanotube shells (sample from Table 1, no. 3); micrographs taken on a JEOL 2010 transmission electron microscope operating at an accelerating voltage of 200 kV (scale bar: 5 nm).

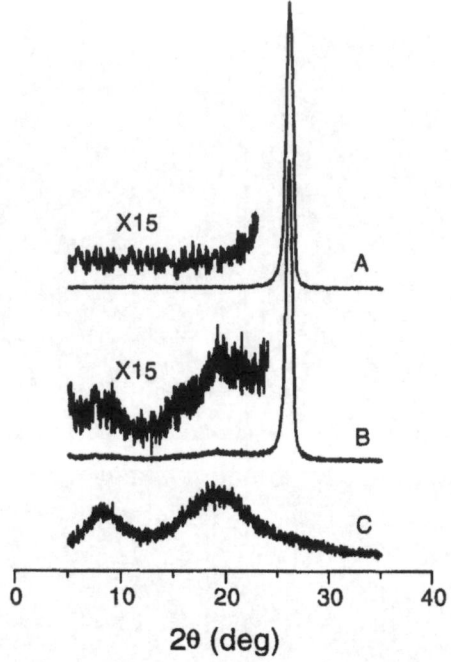

Figure 5 X-ray diffraction patterns of (A) carbon nanotubes, (B) PPA-wrapped carbon nanotubes (sample from Table 1, no. 3), and (C) PPA; diffratograms recorded on a powder diffractometer (Philips PW1830) with Cu $K\alpha$ radiation.

To check whether or not the PPA chains can be attached to the nanotubes by simple physical blending, a control experiment was conducted, in which the nanotubes and the preformed (pure) PPA were admixed and stirred in toluene for 24 hours. Following the same filtration and precipitation procedures detailed in the Experimental Section, all the nanotubes were recovered and the colour of the PPA solution remained unchanged. The microscopic analyses show neither the PPA chains in the recovered nanotubes nor the tubular entity in the recovered PPA. The nanotubes thus may be wrapped by the propagating species of the PPA chains *in situ* during the polymerisation reactions.

Terminal acetylenes (RC≡C-H) often form ≡C-H···π hydrogen bonds with molecules and/or groups that are rich in π-electrons (Kong and Tang, 1998; Stein, 1995; Tang, *et al.*, 1998; Weiss, 1997). Because the nanotubes are full of π-electrons, the PA monomers are likely to "wet" the surfaces of the tubule shells through the ≡C-H···π hydrogen bonds, and polymerisation of such absorbed PA monomers would produce PPA chains wrapping up the carbon nanotubes. Polymerisations of internal acetylenes without the acidic acetylene hydrogen such as CH$_3$C≡CPh fail to incorporate any nanotubes into the resulting polyacetylenes, further suggesting the important role of the ≡C-H···π hydrogen bonds *in the* wrapping processes of the nanotubes by the PPA chains. The *trans*-cisoidal and *trans*-transoidal PPAs can form helices with 3–12 monomer units per coil (Furlani *et al.*, 1986), polyacetylenes with chiral substituents possess helical conformation (Aoki *et al.*, 1993; Ciardeli *et al.*, 1974; Kishimoto *et al.*, 1995; Moor *et al.*, 1991; Tang, 1990; Tang and Kotera, 1989), and complexation of PPA derivatives with chiral molecules can induce helix formation (Yashima *et al.*, 1998). The nanotubes are cylinders of rolled-up graphene sheets with various chiralities, and the propagating species of the polymer chains may experience steric and electronic interaction with the chiral tubules, thus generating the PPA helices spiralling around the nanotubes. When the short nanotubes are thickly wrapped by the PPA chains, the solvation of the polymer molecules may drag the tubules into the solvent, thus making the nanotubes "soluble". The principle for the dissolution of the wrapped short nanotubes in organic solvents is similar to that for the solubilization of fullerene clusters in polar solvents such as trifluoroacetic acid in the "microcontainer" system recently developed by Jenekhe and Chen (1998). On the other hand, the long tubules unwrapped or thinly wrapped by the PPA chains would remain insoluble in the organic solvents.

The solubility of the short nanotubes wrapped by the PPA chains enables the structural characterisation by wet spectroscopy and the investigation of their solution properties. The ^{13}C nuclear magnetic resonance (NMR) spectrum of the chloroform solution of the NT–PPA is almost identical to that of the parent PPA, except for that the peak of the NT–PPA at δ *ca.* 140 becomes more intense. The solid-state NMR spectrum of the parent nanotubes shows a broad peak centred at δ 130 and the acid-functionalised nanotubes absorb at δ *ca.* 145 (Rao *et al.*, 1996). The increase in the peak intensity at δ *ca.* 140 thus may be attributed to the absorption of the nanotubes wrapped by the PPA chains.

3.2 Shear-Induced Alignments

We have recently found that liquid crystalline polyacetylenes can be oriented by external forces (Kong and Tang, 1998; Tang *et al.*, 1998), and it is of interest to know whether the nanotubes in the NT–PPA solutions can be aligned by macroscopic processing means such as mechanical shear. We thus applied a shear force to a thin layer of concentrated THF solution of NT–PPA sandwiched between two pieces of clean glass slides. After a while, the initially uniform background of the homogeneous solution was found to be embellished with numerous parallel spots aligning along the shear direction (Figure 6A). Such texture is not observable in the control experiment using THF solution of pure PPA, implying that the tiny spots are the aggregates of the nanotubes in the NT–PPA. As the solvent of the sheared NT–PPA solution in the glass cell slowly evaporates, the nanotubes oriented by the shear force would gradually aggregate. The surrounding PPA chains, however, would prevent the aggregates from growing bigger, thus resulting in the formation of the tiny micelle-like spots. During the emerging stage of the aggregates from the homogeneous solution, necklace-like texture forms, in which the tiny black "pearls" of the nanotube clusters are strung together, probably by the oriented PPA chains (Figure 6B). This suggests that the alignment of the nanotube aggregates along the shear direction is assisted, at least in part, by the shear-induced orientation of the stiff polyacetylene chains. When a curved or bent shear force is applied, the nanotubes align along the curvature of shear direction (Figure 6C), demonstrating the ease with which the nanotubes may be manipulated and aligned by simple mechanical perturbation.

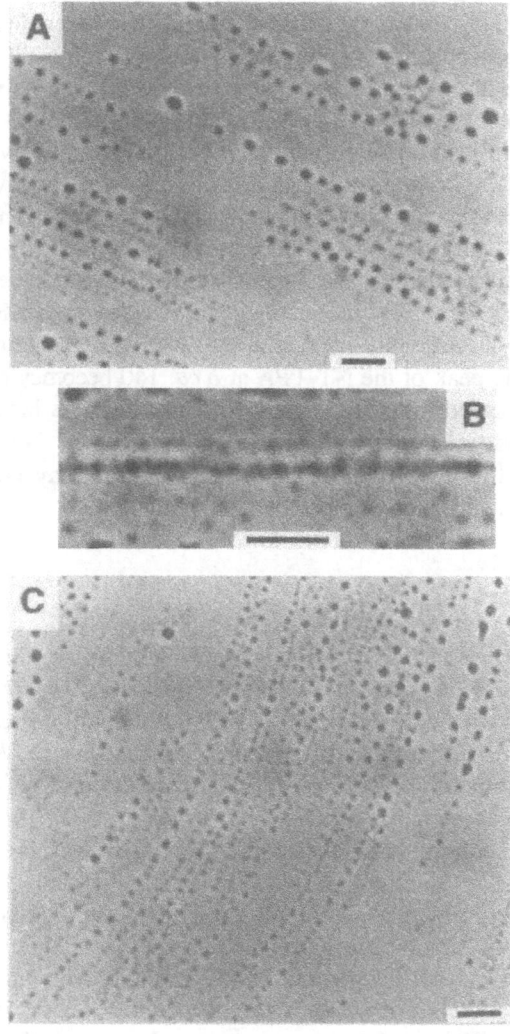

Figure 6 Alignment of carbon nanotubes in NT–PPA induced by mechanical shear; photomicrographs taken on an Olympus BX60 optical microscope (scale bar: 10 μm).

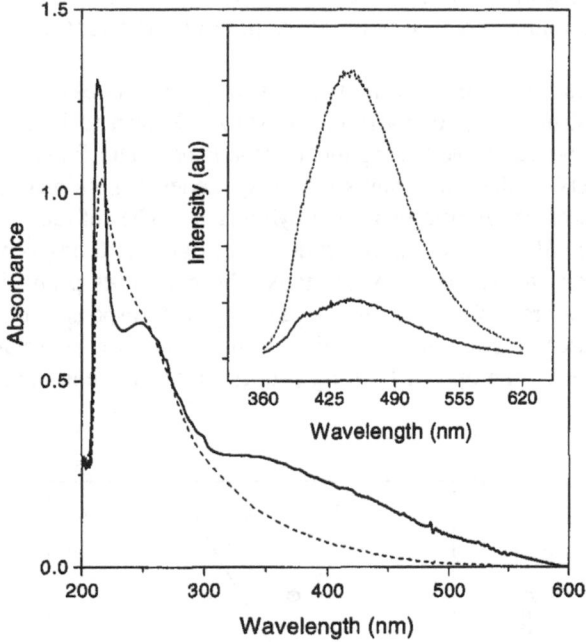

Figure 7 Electronic absorption and fluorescence emission (inset) spectra of THF solutions of NT–PPA (solid line; sample from Table 1, no. 1) and PPA (dotted line) recorded respectively on a Milton Roy 3000 Array spectrophotometer and a SLM Aminco JD-490 spectrofluorometer (excited at 350 nm); concentration (mg/mL): 0.01 (absorption), 0.10 (emission).

3.3 Electronic and Optical Properties

While the parent PPA weakly absorbs in the visible spectral region, the absorption of the NT–PPA well extends to *ca.* 600 nm, in agreement with its deeper colour (Figure 7). When the PPA is excited at 350 nm, it emits fluorescence with a peak maximum at *ca.* 450 nm (Lee *et al.*, 1999). The fluorescence spectral profile of the NT–PPA is similar to that of the PPA, but with much lower quantum yield. Curran *et al.* (1998) have recently observed similar phenomenon in their nanotube–poly(*m*-phenylenevinylene-*co*-2,5-dioctoxy-*p*-phenylenevinylene) (NT–PPV) composite system. The fluorescence of the NT–PPV was much weaker than that of the parent polymer, "caused", as the authors proposed, "by absorption, quenching, and scattering from the nanotubes present". The similar mechanisms may be applicable to the decrease in the fluorescence intensity in our NT–PPA system. The intimate wrapping of the polyacetylene chains around the nanotube shells may be viewed as a perfect donor-acceptor heterojunction (Yu *et al.*, 1995). Electrons in the excited PPA chains may move to the nanotube shells, and the photoinduced charge transfer may thus effectively quench the photoluminescence of the PPA molecules (Saricifti *et al.*, 1992; Yu *et al.*, 1998).

Fullerenes (Tutt and Kost, 1992) and polyacene-based oligomers with graphite-like structures (Kojima *et al.*, 1995) are known to limit intense optical pulses by reverse saturable absorption mechanisms. The fullerene tips and the graphene sheets of the short nanotubes may undergo nonlinear optical (NLO) absorption processes. Furthermore, the cylindrical bodies of the nanotubes, albeit with high length/diameter ratio, may function as light scattering centres. Both the NLO absorption and scattering would make the nanotubes promising candidates for optical limiters. We thus investigated optical responses of the NT–PPA solutions to laser pulses, employing the similar experimental setup used in our previous studies on the optical limiting properties of fullerene materials (Tang *et al.*, 1998).

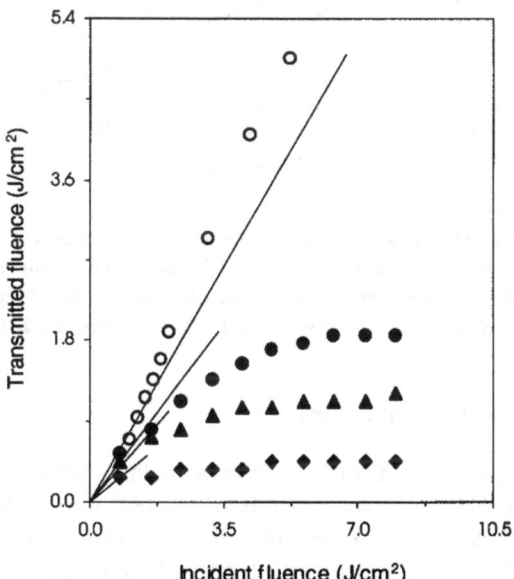

Figure 8 Optical limiting responses of THF solutions of the NT–PPA prepared by WCl_6–Ph_4Sn (sample from Table 1, no. 1); concentration (c; mg/mL)/linear transmittance (T; %): 0.4/57 (●), 0.5/48 (▲), 0.6/34 (◆). Optical responses of a THF solution of PPA are shown for comparison [c (mg/mL)/T (%): 4.0/75 (○)]. The 8-ns pulses of 532-nm light were generated from a Quanta Ray GCR 3 frequency-doubled Q-switched Nd:YAG laser.

When a THF solution of the parent PPA is shot by the 8-ns pulses of 532-nm laser light, the transmitted fluence linearly increases in the region of low incident fluence (linear transmittance: 75%; Figure 8). The output starts and continues to deviate from the linear-transmission line from the input of *ca.* 1.7 J/cm^2, implying that the intense illumination gradually bleaches the PPA to transparency, probably by the laser-induced photolysis (degradation) of the polyacetylene chains. The NT–PPA solutions, however, respond to the optical pulses in a strikingly different way. The linear transmittance of a dilute NT–PPA solution (0.4 mg/mL) is only 57% (Figure 8), although its concentration is only one tenth of that (4 mg/mL) of the parent PPA, probably because of the optical losses caused by the nanotube absorption and scattering. As the incident fluence increases, the NT–PPA solution becomes opaque, instead of transparent, with its transmitted fluence eventually levelling off or saturating at 1.85 J/cm² (saturation fluence). Clearly the nanotubes have endowed the NT–PPA with optical limiting power. The energy-sinking and radical-trapping functions of aromatic rings often protect polymer molecules from photodegradation (Allen and Edge, 1992; Hamid *et al.*, 1990; Scott, 1990), and the extensively conjugated graphitic aromatic system of the nanotubes may have enhanced the resistance of the PPA chains against the laser irradiation. As the concentration of the NT–PPA solution increases, its saturation fluence decreases. Increasing the concentration to 0.6 mg/mL readily decreases the saturation fluence to as low as 0.45 J/cm².

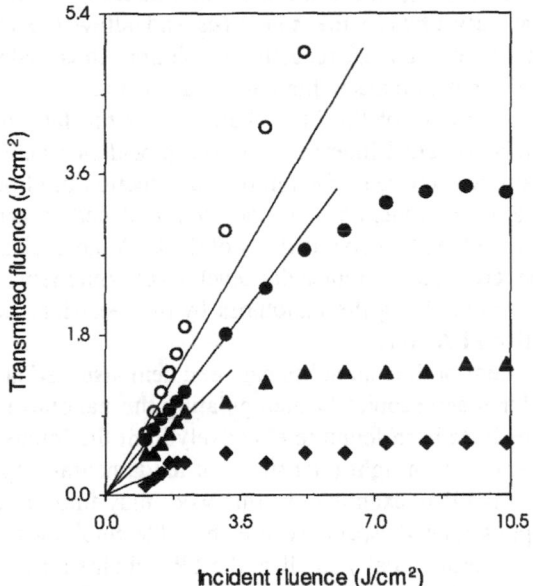

Figure 9 Optical limiting responses of THF solutions of the NT–PPA prepared by [Rh(nbd)Cl]₂ (sample from Table 1, no. 4); concentration (*c*; mg/mL)/linear transmittance (*T*; %): 1.0/57 (●), 2.0/42 (▲), 3.0/19 (◆). Optical responses of a THF solution of PPA are shown for comparison [*c* (mg/mL)/*T* (%): 4.0/75 (○)]. The 8-ns pulses of 532-nm light were generated from a Quanta Ray GCR 3 frequency-doubled Q-switched Nd:YAG laser.

Similarly, the NT–PPA prepared by the [Rh(nbd)Cl]$_2$ catalyst also limits the intense optical pulses (Figure 9). A THF solution of the NT–PPA with a concentration of 1.0 mg/mL exhibits a saturation fluence of 3.44 mg/cm^2, which is much higher than those of the THF solutions of the NT–PPA prepared by WCl$_6$–Ph$_4$Sn with lower concentrations (*cf.* Figure 8). The major difference between the two NT–PPAs is their nanotube contents. The NT–PPA with lower nanotube content shows higher saturation fluence, further confirming that the nanotubes are responsible for the optical limiting in the NT–PPA solutions. In this case again, increasing the concentration decreases the saturation fluence. A low saturation fluence of 0.58 J/cm^2 is achieved when the concentration is increased to 3.0 mg/mL.

4 CONCLUSION

In summary, in this study, short carbon nanotubes have been solubilized by wrapping them with soluble PPA chains. Solubility is the primary requirement for studying wet chemistry, and this work paves the way for new developments in nanotube research. Heating an NT–PPA solution under controlled conditions, for example, will partially unwrap the nanotubes and allow the naked part of the nanotubes to react with chemical reagents in solution, thus making it possible to study chemistry of the nanotubes in homogeneous media.

The helical wrapping of the PPA chains along the latitudes of the carbon nanotubes is of technological interest. The acceptor-donor heterojunction of the NT–PPA, for example, may be utilised in the construction of photovoltaic devices. The charge separation generated by the photoinduced charge transfer may dramatically increase the photoconductivity of the PPA molecules. One intriguing possibility is the creation of molecular-level electromagnetic devices such as "nanomotors" by magnetising the nanotubes by the electrical field generated by the photoconductive PPA coils.

Ready alignment of the nanotubes has been demonstrated and this provides a versatile means for macroscopically manipulating the nanotubes. Moreover, the NT–PPA solutions have been found to effectively limit the intense 532-nm optical pulses. Since the control of light intensity is of fundamental importance in optics engineering, the NT–PPAs examined in this work may find an array of potential applications in optics-related especially laser-based technologies.

The nanotubes dramatically stabilise the PPA chains against the harsh laser irradiation. Such stabilisation effect is of academic and practical significance, which will not only trigger basic research on the understanding of electronic properties of the nanotubes in solutions by wet spectroscopy, but also help find technological applications for the nanotubes in electronic and optical systems. For example, incorporation of the nanotubes into optical-limiting and light-emitting polymers would greatly enhance their lifetime, which has been a thorny problem for the devices constructed from the organic materials.

5 ACKLOWLEDGEMENT

This work was in part supported by the Research Grants Council of the Hong Kong Special Administrative Region, China (Project No.: HKUST6062/98P) and by the Joint Laboratory for Nanostructured Materials and Technology between the Chinese Academy of Sciences and the Hong Kong University of Science & Technology (HKUST). We thank Ms. Y. Zhang of our Materials Characterisation & Preparation Facility for her help in the SEM and TEM measurements and Mr. M. W. Fok of our Chemistry Department for his assistance in the optical limiting experiment.

6 REFERENCES

Ajayan, P.M., Ebbesen, T.W., Ichihashi, T., Iijima, S., Tanigaki, K. and Hiura, H., 1993, *Nature*, **362**, 522.
Akagi, A. and Shirakawa, H., 1996, *Macromolecular Symposium*, **104**, 137.
Allen, N.S. and Edge, M., 1992, *Fundamentals of Polymer Degradation and Stabilisation*, (Elsevier Applied Science: London).
Aoki, T., Kokai, M., Shinohara, K. and Oikawa, E., 1993, *Chemistry Letters*, 2009.
Ciardelli, F., Lanzillo, S. and Pieroni, O., 1974, *Macromolecules*, **7**, 174.
Curran, S., Ajayan, P.M., Blau, W.J., Carroll, D.L., Coleman, J.N., Dalton, A.B., Davey, A.P., Drury, A., Drury, A., McCarthy, B., Meier, S. and Strevens, A., 1998, *Advanced Materials*, **10**, 1091.
Ebbesen T.W., 1997, *Carbon Nanotubes: Preparation and Properties*, (CRC Press: Boca Raton, FL).
Endo, M., Iijima, S. and Dresselhaus, M.S., 1996, *Carbon Nanotubes*, (Pergamon Press: Oxford).
Falvo. M.R., Clary, G.J., Taylor, R.M., II., Chi, V., Brooks, F.P., Jr. and Superfine, R., 1997, *Nature*, **389**, 582.
Furlani, A., Napoletano, C., Russo, M.V. and Feast, W.J., 1986, *Polymer Bulletin*, **16**, 311.
Hamid, S.H., Amin, M.B. and Maadhah, A.G., 1992, *Handbook of Polymer Degradation*, (Dekker: New York).
Jenekhe, S.A. and Chen, X.L., 1998, *Science*, **279**, 1903.
Kang, E.T., Ehrlich, P., Bhatt, A.P. and Anderson, W.A., 1984, *Macromolecules*, **17**, 1020.
Kishimoto, Y., Itou, M., Miyatake, T., Ikariya, T. and Noyori, R., 1995, *Macromolecules*, **28**, 6662.
Kojima, Y., Matsuoka, T., Sato, N. and Takahashi, H., 1995, *Macromolecules*, **28**, 2893.

Kong, X. and Tang, B.Z., 1998, *Chemistry of Materials*, **10**, 3352.

Lee, C.W., Wong, K.S. and Tang, B.Z., 1999, *Synthetic Metals*, in press.

Masuda, T., Tang, B.Z., Higashimura, H. and Yamaoka, H., 1985, *Macromolecules*, **18**, 2369.

Moore, J.S., Gorman, C.B. and Grubbs, R.H., 1991, *Journal of American Chemical Society*, **113**, 1704.

Odom, T.W., Huang, J.-L, Kim, P. and Lieber, C.M., 1998, *Nature*, **391**, 62.

Peng, H., Leung, S.M. and Tang, B.Z., 1997, *Chinese Journal of Polymer Science*, **15**, 193.

Rao, C.N.R., Govindaraj, A. and Satishkumar, B.C., 1996, *Chemical Communications*, 1525.

Sariciftci, N.S., Smilowitz, L., Heeger, A.J. and Wudl, F. 1992, *Science*, 1474.

Scott, G., 1990, *Mechanisms of Polymer Degradation and Stabilisation*, (Elsevier: London).

Stein, T., 1995, *Chemical Communications*, 95.

Taliani, C., Vardeny, Z.V. and Maruyama, Y., 1993, *Synthetic Metals for Non-Linear Optics and Electronics*, (European Materials Research Society: Amsterdam).

Tang, B.Z., 1990, *Advanced Materials*, **2**, 107.

Tang, B.Z., 1996, *Advance Materials*, **8**, 939.

Tang, B.Z., 1998, *Proceedings of IUPAC 9th International Symposium on Novel Aromatic Compounds*, IL15.

Tang, B. Z., 1998, *Proceedings of the IUPAC World Polymer Congress Macro'98*, 772.

Tang, B.Z., Kong, X., Wan, X., Peng, H., Lam, W.Y., Feng, X.-D. and Kwok, H.S., 1998, *Macromolecules*, **31**, 2419.

Tang, B.Z. and Kotera, N., 1989, *Macromolecules*, **22**, 4388.

Tang, B.Z., Leung, S.M., Peng, H., Yu, N.T. and Su, K.C., 1997, *Macromolecules*, **30**, 2848.

Tang, B.Z. and Peng, H., 1997, In *Recent Advances in Overseas Polymer Research*, Chapter 10, edited by He, T. and Hu, H. (Chemical Industry Press: Beijing), 165-173.

Tang, B.Z., Peng, H., Leung, S.M., Au, C.F., Poon, W.H., Chen, H., Wu, X., Fok, M.W., Yu, N.-T., Hiraoka, H., Song, C., Fu, J., Ge, W., Wong, K.L.G., Monde, T., Nemoto, F. and Su, K.C., 1998, *Macromolecules*, **31**, 103.

Tang, B.Z., Peng, H., Leung, S.M., Yu, N.-T., Hiraoka, H. and Fok, M.W., 1998, In *Materials for Optical Limiting II*, edited by Sutherland, R., Pachter, R., Hood, P., Hagan, D., Lewis, K. and Perry, J. W. (Materials Research Society: Pittsburgh, PA), 69.

Treacy, M.M.J., Ebbesen, T.W. and Gibson, J.M., 1996, *Nature*, **381**, 678.

Tsang, S.C., Harris, P.J.F. and Green, M.H.L., 1993, *Nature*, **362**, 520.

Tutt, L.W. and Kost, A., 1992, *Nature*, **356**, 225.

Weiss, H.-C., Blaser, D., Boese, R., Dougham, B.M. and Haley, M.M., 1997, *Chemical Communications*, 1703.

Weiss, H.-C., Boese, R., Smith, H.L. and Haley, M.M., 1997, *Chemical Communications*, 2403.

Wildoer, J.W.G., Venema, L.C., Rinzler, A.G., Smalley, R.E. and Dekker, C., 1998, *Nature*, **391**, 59.

Xu, H. and Tang, B. Z., 1998, *Abstract of the Fifth Symposium on Chemistry Postgraduate Research in Hong Kong*, 28.

Yashima, E., Maeda, Y. and Okamoto, Y., 1998, *Journal of American Chemical Society*, **120**, 8895.

Yu, G., Gao, J., Hummelen, J.C., Wudl, F. and Heeger, A. J., 1995, *Science*, 1789.

Yu, G., Wang, J., McElvain, J. and Heeger, A.J., 1998, *Advanced Materials*, 1431.

14 Production of Very Long Aligned Carbon Nanotubes

Z. W. Pan and S. S. Xie
Institute of Physics, Chinese Academy of Sciences, P. O. Box 603, Beijing 100080, China

ABSTRACT

Very long carbon nanotubes with length of about 2 mm has been prepared by pyrolysis of acetylene over film-like iron/silica substrates. The nanotubes with diameter of 20–40 nm grew outwards perpendicularly from the surface of the substrates and formed aligned array of isolated tubes with spacing between the tubes of about 100 nm. The mechanical properties of very long nanotubes were measured by pulling nanotube ropes with a special designed stress-strain puller.

1. INTRODUCTION

Carbon nanotubes have been intensively studies since their discovery (Iijima, 1991). Nanotubes can now be produced in large quantities by either arc discharge (Ebbesen and Ajayan, 1992) or thermal decomposition of hydrocarbon (Li *et al.*, 1996). Controlled growth of aligned carbon nanotubes is important for both fundamental research and technological application. Several groups have been successfully prepared aligned carbon nanotubes on different substrates by chemical vapor deposition (CVD) (Li *et al.*, 1996; Terrones *et al.*, 1997; Ren *et al.*, 1998). However, the nanotubes prepared are too short ($\leq$ 100 µm) to be evaluated by regular methods. We report here the growth of very long ($\sim$ 2 mm) aligned carbon nanotubes by pyrolysis of acetylene over film-like iron/silica substrates.

Using the very long carbon nanotubes, we measured the mechanical properties including Young's modulus and tensile strength of nanotubes with a stress-strain puller.

2. EXPERIMENTAL

The film-like iron/silica substrates were prepared by a sol-gel process from tetraethoxysilane hydrolysis in iron nitrate aqueous solution, which has been described in detail elsewhere (Pan *et al.*, 1998). During pyrolysis, carbon nanotubes were formed on the substrates by decomposition of acetylene at 600 °C. The growth time varied from 1 to 48 hours.

The as-formed nanotubes were analyzed by scanning electron microscope (SEM) and energy dispersive x-ray (XRD) attached to the SEM. A transmission electron microscope (TEM) was used to characterize the finer structure of tubes.

For mechanical property measurement, a special designed stress-strain puller was used. This puller can apply an axial force (uniaxial stress) to a rope containing several ten thousands nanotubes and simultaneously measure the corresponding rope elongation. The diagram and measurement method of this device has been described in detail by Stove *et al.* (1991).

3.   RESULT AND DISCUSSION

We have used this method to prepare nanotubes at a very high yield. Low-magnification SEM images show that every surface of the substrate is covered with a nanotube array composed of large quantities of highly aligned carbon nanotubes

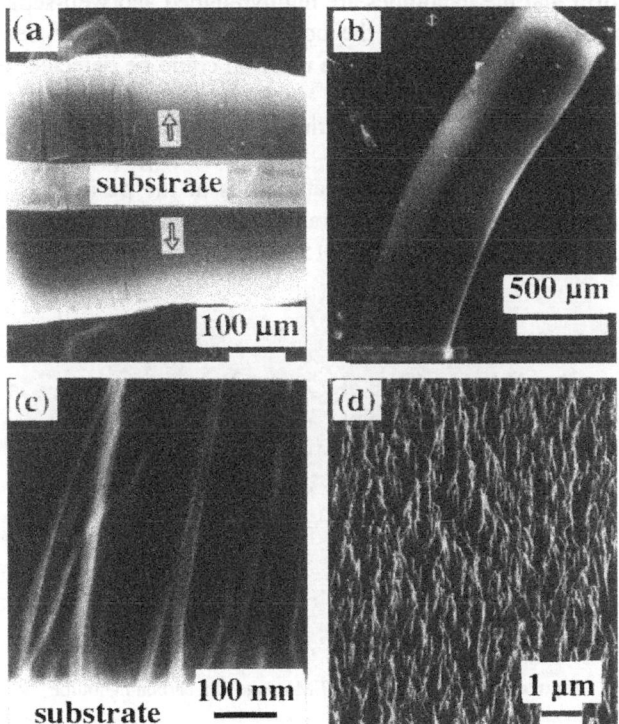

Figure 1   SEM images of aligned carbon nanotube arrays. (a) A sample after 5 hours of growth. The growth direction of nanotubes are labeled by arrows. (b) A sample after 48 hours of growth. The length of the nanotube array reaches ~ 2 mm. (c) High-magnification SEM image of an aligned nanotube array. (d) The SEM image of the bottom end of an aligned nanotube array.

(Figure 1a). The length of the nanotubes increase with the growth time, the growth rate being about 30–40 μm/hour. After 48 hours growth, the length of the nanotubes reaches ~ 2 mm (Figure 1b), which is an order of magnitude longer (1 mm vs. 100 μm) than that described in most previous reports. We believe that the nanotubes could be even longer if the growth time was further increased. The area of every aligned nanotube array is equal to the relevant surface area of the substrate. Using a modified sol-gel process, we can produce substrate with large flat surface and thus synthesize large area of aligned carbon nanotubes. The largest area we have obtained is up to 1 cm^2.

High-magnification SEM studies reveal that carbon nanotubes grow outwards perpendicularly and separately from the substrate and form an aligned nanotube array (Figure 1c). The nanotubes within the array have uniform external diameter (20–40 nm) and spacing (~100 nm) between tubes. We note that no traces of polyhedral particles or other graphitic nanostructures are detected in the array, which indicates that the nanotubes prepared in this study have very high purity and thus have very high quality.

The nanotube array can be easily stripped off from the substrate without destroying the array's integrity, and the SEM image taken from the bottom end of the array confirm that the nanotubes are highly aligned and well-separated (Figure 1d). EDX spectra collected from the bottom end of the array demonstrate the presence of carbon alone, neither silicon nor iron could be detected, which further confirms the high purity of our sample.

High-resolution TEM observations show that the nanotubes are well graphitized and typically consist of 10 ~ 30 concentric graphite layers (Figure 2). Only small amount of carbonaceous materials is at the periphery of the carbon nanotubes due to the low growth temperature (600 °C). The inner diameter of the carbon nanotubes is in the range of 10 ~ 15 nm.

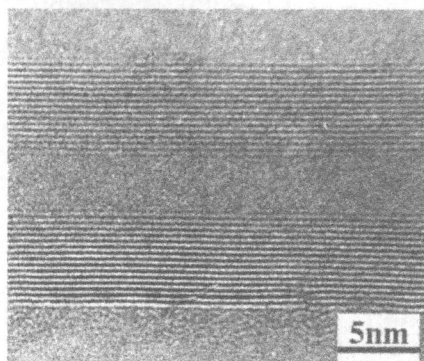

5nm

Figure 2   High-resolution TEM image of a carbon nanotube

The successful preparation of very long aligned carbon nanotubes make us possible to manipulate the tubes under a light microscope and measure some of their properties by regular methods. For example, we can cleave a long rope containing ten thousands nanotubes from the array along the growth direction of the nanotubes, and mount it on the two grips of a special designed stress-strain puller. The force-displacement data is recorded during pulling the nanotube rope. The Young's modulus of the nanotubes can be given by the slope of a stress-strain curve: $Y = \Delta\sigma/\Delta\varepsilon$, where σ is the uniaxial stress and ε is the strain. The tensile strength can be determined by the maximum uniaxial stress. The average Young's modulus and tensile strength we obtained are 0.45 ± 0.23 TPa and 1.72 ± 0.64 GPa, respectively. Although these values are much lower than those measured for nanotubes produced by arc discharge method (Treacy *et al.*, 1996; Wong *et al.*, 1997), we believe that they are reasonable for nanotubes produced by CVD method, since CVD nanotubes have a much greater number of defects than tubes produced by arc method. These defects can severely limit the mechanical strength of nanotubes.

1. CONCLUSIONS

We have prepared very long (~ 2 mm) aligned carbon nanotubes by pyrolysis of acetylene over iron/silica substrates at 600 °C for 48 hours. Using the very long carbon nanotubes, we measured the mechanical properties of nanotubes and found the average Young's modulus and tensile strength are 0.45 ± 0.23 TPa and 1.72 ± 0.64 GPa, respectively.

REFERENCES

Ebbesen, T.W. and Ajayan, P.M., 1992, Large-scale synthesis of carbon nanotubes. *Nature*, **358**, pp. 220–222.

Iijima, S., 1991, Helical microtubules of graphitic carbon. *Nature*, **354**, pp. 56–58.

Li, W.Z., *et al.*, 1996, Large-scale synthesis of aligned carbon nanotubes. *Science*, **274**, pp. 1701–1703.

Pan, Z.W., *et al.*, 1998, Very long carbon nanotubes. *Nature*, **394**, pp. 631–632.

Ren, Z.F., *et al.*, 1998, Synthesis of large arrays of well-aligned carbon nanotubes on glass. *Science*, **282**, pp. 1105–1107.

Skove, M.J., *et al.*, 1991, Device for simultaneously measuring stress, strain, and resistance in "whiskerlike" materials in the temperature range $1.5K<T<360K$. *Review of Scientific Instrument*, **62**, pp. 1010⁻1014.

Terrones, M., *et al.*, 1997, Controlled production of aligned nanotube bundles. *Nature*, **388**, pp. 52–55.

Treacy, M.M.J., Ebbesen, T,W. and Gibson, J.M., 1996, Exceptionally high Young's modulus observed for individual carbon nanotubes. *Nature*, **381**, pp. 678–680.

Wong, E.W., sheehan, P.E. and Lieber, C.M., 1997, Nanobeam mechanics: elasticity, strength, and toughness of nanorods and nanotubes. *Science*, **277**, pp. 1971–1975.

15 Electronic Structure of the Endohedral Metallofullerene Lu@C$_{82}$

Houjin Huang and Shihe Yang
Department of Chemistry, Hong Kong University of Science and Technology, Clear Water Bay, Hong Kong

ABSTRACT

XPS, EPR and UV-Vis-Nir spectroscopic techniques are employed to characterize the mono-metallofullerene Lu@C$_{82}$. Our results suggest that the encaged Lu atom donates approximately three electrons to the carbon cage.

Introduction

The notion that the central cavity of a fullerene can trap atoms or molecules has met with great fascination since the beginning of the fullerene discovery (Heath, 1986). Fullerene is a perfectly engineered nanoparticle with a unique carbon surface structure and a hollow interior. It is self-stabilized with no need for capping ligands or polymers. It is expected that the incorporation of metal atoms into the fullerene cages would significantly modify the fullerene electronic structure, providing a novel type of materials with a host of applications in superconductors, lasers, ferroelectrics, etc.

There has been keen interest in the electronic structure of metallofullerenes. Experimental and theoretical studies have contributed significantly to the understanding of this issue. It is generally accepted both divalent and trivalent lanthanide ions can be trapped inside the fullerene cage for M@C$_{82}$. In this work, we concentrate on XPS, EPR and UV-Vis-Nir studies of the metallofullerene Lu@C$_{82}$, which has relatively paucity of data. What is interesting about Lu@C$_{82}$ is that the 4f shell of Lu is fully closed, significantly simplifying the interpretation of the spectroscopic data of the metallofullerene. A theoretical study (Kobayashi, 1998) predicted that in Lu@C$_{82}$, one 5d electron and one 6s electron are transferred, with the remaining one 6s electron largely on Lu because of the large relativistic contraction and stabilization of the 6s orbital. If this is the case, a large hyperfine coupling constant is expected in EPR, which differs greatly from those of M@C$_{82}$ (M = Sc, Y, La) with a slight s spin density on M.

Experiment

Carbon soot containing metallofullerenes was produced by the standard arc vaporization method using a composite anode under an He atmosphere of 125 Torr. The composite anode contained graphite powder and lanthanide metal oxides in an atomic ratio metal/C $\approx$ 0.02.

The raw soot was collected and extracted for 24 h using DMF as the solvent (Ding, 1996). The extract was filtered using a slow-rate filter paper and a brownish green solution was obtained. After removal of DMF by vacuum evaporation, a black powder was collected and redissolved in toluene. The solution was filtered with 0.2 μM disk filter (Rubbermaid Inc.) before HPLC separation. For HPLC separation, a PYE Cosmosil column (10 mm $\times$ 250 mm, Nacalai Tesque Inc.) was employed with toluene being the mobile phase. The injection volume was 5 mL and the elution rate was 4.0 mL/min.

DCI negative ion mass spectrometry (Finnigan TSQ 7000) was used to characterize the composition of the samples. For XPS study, the gold $4f_{7/2}$ line at 83.9 eV was used as the reference for the measurement of binding energies. The EPR spectra were recorded using a JOEL JES200 spectrometer equipped with a liquid nitrogen cryostat. An X-band frequency of 9.437 GHz was used with a field modulation frequency of 100 kHz. Before measurements, the solution of Lu@C$_{82}$ was degassed by a freeze-pump-thaw cycle. Mn^{2+} in MgO was used as an EPR marker to calibrate the g values. The UV-Vis-Nir absorption spectrum of the metallofullerene in toluene solution was obtained with a Perkin-Elmer spectrometer (Lambda 19).

Results and discussion

XPS is a powerful technique for estimating the oxidation state of Lu inside the fullerene cage. The 4f shell of Lu is fully occupied and shake-down peaks due to charge transfer from the fullerene cage to the 4f orbitals of Lu can be excluded. The valence band spectrum of Lu is easily obtainable, which is supposed to be very sensitive to the valence state of the metal atom. Fig. 1 shows the XPS spectra of Lu@C$_{82}$ and LuCl$_3$ in the 4f region. The two distinctive peaks are due to the spin-orbit splitting: $4f_{7/2}$ and $4f_{5/2}$ at 9.3 and 10.7 eV, respectively. It is clear that the two spectra are very similar in terms of the peak splitting, the peak ratio of the two peaks, and the locations. The only difference is that the peaks of LuCl$_3$ are somewhat broader due to the poor conductivity of th sample. No other peaks have been observed around this spectral region aside from the doublet. Since it is generally accepted that Lu in LuCl$_3$ has an oxidation state of 3+, the similarity of the XPS spectra shows that the Lu atom has donated three electrons to the carbon cage C$_{82}$. Furthermore, the XPS spectrum of Lu@C$_{82}$ in the 4d region is also similar to that of LuCl$_3$. Although this spectral region of is close to Cl 2p which possibly contaminating the spectra, meaningful comparisons can still be made. Nagase et al. predicted that around one electron is left in the 6s orbital of Lu in Lu@C$_{82}$ due to the inertness of the 6s orbital. However, significantly changed XPS features would

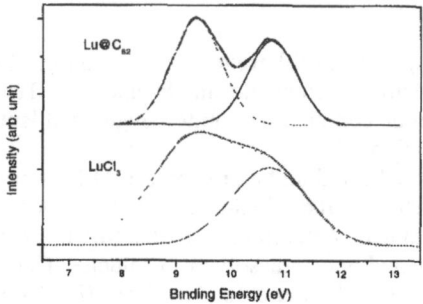

Figure 1 XPS spectra of Lu@C$_{82}$ and LuCl$_3$ in the 4f region of Lu.

be expected if the 6s orbital of Lu has one electron left, which we have not observed.

    Shown in Fig. 2 are the EPR spectra of Lu@C$_{82}$ at room temperature and at liquid nitrogen temperature. Toluene was used as the solvent. A single broad peak is observed at both room temperature and liquid nitrogen temperature. Since for the nuclear spin of ^{175}Lu, I = 7/2, eight equally-spaced hyperfine lines are expected. These hyperfine features could not be resolved. The g value at room temperature is 1.99705 while at liquid nitrogen temperature it is 2.00085. At room temperature, the broad spectrum exhibits a characteristic Lorentzian line shape. However, at liquid nitrogen temperature, the spectral shape is deformed with the low field part of the spectrum being more intense. Presumably, this is due to the motional effects of anisotropic interactions of the spin system, leading to an m$_I$ dependence of the peak widths. It is reasonable that the spin relaxation processes are controlled at liquid temperature by the viscosity of the solvent whereas at room temperature an intrinsic process is dominant. Knapp et al. (Knapp, 1998) found that at temperatures above 250 K, a thermally activated process, leads to pronounced line broadening. Clearly, the unpaired electron has to be from the carbon cage because a much larger hyperfine coupling constant is expected if it is in the 6s orbital of Lu.

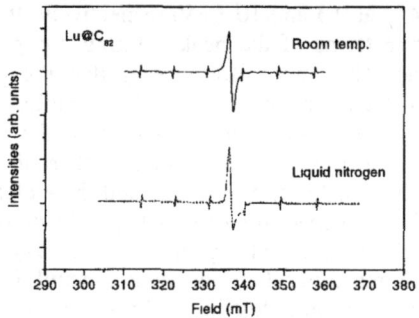

Figure 2 EPR spectra of Lu@C$_{82}$ at room temperature and liquid nitrogen temperature.

Figure 3 shows the UV-Vis-Nir absorption spectrum of $Lu@C_{82}$ in toluene. This spectrum is very similar to those of other monometallofullerenes $M@C_{82}$ (M = rare earth metal atoms) (Ding, 1997). Three prominent absorption features are observed at around 630, 1000 and 1410 nm (broad). The similarity of the absorption spectra of the metallofullerene family in terms of the almost exact match of peak locations suggests that the spectral features are derived from the carbon cage with an open-shell electronic structure and that the cage accepted roughly the same amount of electrons from the metals, i.e., they may be represented by an electronic structure of the form $M^{3+}@C_{82}^{3-}$.

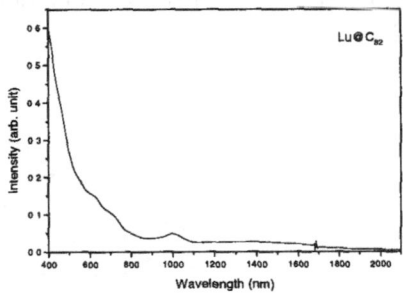

Figure 3 UV-Vis-Nir absorption spectrum of $Lu@C_{82}$ in toluene.

Summary and conclusions

In summary, the metallofullerene $Lu@C_{82}$ has been characterized by XPS, EPR and UV-Vis-Nir spectroscopic techniques. All the spectroscopic evidence suggests that the lanthanide metal atom Lu in $Lu@C_{82}$ displays a trivalent state in contrast to the prediction based on ab initio calculations. This represents a great challenge for the theoretical investigations.

References

Heath, J.R., O'Brien, S.C., Zhang, Q., Liu, Y., Curl, R.F., Kroto, H.W., Tittel, F.K. and Smalley, R.E., 1985, Lanthanum complexes of spheroidal carbon shells. *Journal American Chemical Society*, **107**, pp. 7779-7780.

Kobayashi, K. and Nagase, S., 1998, Structures and electronic states of $M@C_{82}$ (M = Sc, Y, La and lanthanides). *Chemical Physics Letters*, **282**, pp. 325-329.

Ding, J.Q. and Yang, S.H., 1996, Efficient DMF extraction of endohedral metallofullerenes for HPLC purification. *Chemistry of Materials*, **8**, pp. 2824-2827.

Knapp, C., Weiden, N. and Dinse, K.-P., 1998, EPR investigation of endofullerenes in solution. *Applied Physics A*, **66**, pp. 249-255.

Ding, J.Q. and Yang, S.H., 1997, Systematic isolation of endohedral fullerenes containing lanthanide atoms and their characterization. *Journal of Physics and Chemistry of Solids*, **58**, pp. 1661-1667.

16 Localized-Density-Matrix Method, and its Application to Carbon Nanotubes

Satoshi Yokojima, WanZhen Liang, XiuJun Wang, ManFai Ng, DongHao Zhuo and GuanHua Chen
Department of Chemistry, The University of Hong Kong, Pokfulam Road, Hong Kong

ABSTRACT

The localized-density-matrix (LDM) method [Yokojima and Chen, Chem. Phys. Lett. **292**, 379 (1998)] is employed to simulate the optical responses of very large carbon nanotubes. The PPP and PM3 Hamiltonians are used to describe the electrons in the systems, and the time-dependent Hartree-Fock (TDHF) approximation is employed to calculate the linear optical responses. Real space reduced density matrices for different oscillators are examined, and thus the nature of dipole activated excitations is understood.

I.   INTRODUCTION

Carbon nanotubes (Iijima, 1991), which are graphite sheets rolled up in a cylindrical form, are expected to present very interesting optical properties. Carbon nanotubes are quite large, thus are computationally difficult to characterize their excited states properties. Recently the linear scaling LDM method has been developed to evaluate the properties of both ground and excited states (Yokojima and Chen, 1998; Liang *et.al.* 1999; Yokojima and Chen, 1999; Yokojima and Chen, 1999). It is based on the TDHF approximation (Ring and Schuck, 1980) and the truncation of the ground and excited state reduced single-electron density matrices. Its computational time scales linearly with the system size N. The method has been tested successfully to evaluate the optical and ground state properties of conjugated polymers with the orthogonal and non-orthogonal basis sets (Yokojima and Chen, 1998; Liang *et.al.* 1999; Yokojima and Chen, 1999; Yokojima and Chen, 1999). In this work we apply the LDM method to calculate the absorption spectra of a series of carbon nanotubes. In order to understand the nature of various electronic excitations, we examine the reduced density matrices corresponding to these excitations.

II. MODEL

The effective Hamiltonian can be written as

$$H = H_e + H_{ee} + H_{ext}. \tag{1}$$

Here H_e is one-electron part of the Hamiltonian, which is similar to the tight binding model. H_{ee} describes the effective Coulomb repulsion among electrons. H_{ext} represents the effects of the external field. The PPP Hamiltonian considers only the π electron while the PM3 model includes all valence electrons (2s, $2p_x$, $2p_y$, and $2p_z$). H_{ee} is of different form for the PPP and PM3. Details can be found in references (Takahashi and Mukamel, 1994;Stewart, 1989).

III. TDHF AND LDM METHODS

Within the TDHF approximation, the reduced single-electron density matrix satisfies the following equation of motion

$$i\hbar \frac{d}{dt} \rho(t) = [h(t) + f(t), \rho(t)], \tag{2}$$

where $h(t)$ is the Fock matrix, $f(t)$ describes the interaction between an electron and the external field $E(t)$ (Takahashi and Mukamel, 1994). For the linear response of optical properties, its dynamics may be described by the following equation,

$$i\hbar \left(\frac{d}{dt} + \gamma \right) \delta\rho_{ij}^{(1)} = \sum_k (h_{ik}^{(0)} \delta\rho_{kj}^{(1)} - \delta\rho_{ik}^{(1)} h_{kj}^{(0)})$$
$$+ \sum_k (\delta h_{ik}^{(1)} \rho_{kj}^{(0)} - \rho_{ik}^{(0)} \delta h_{kj}^{(1)}) + \sum_k (f_{ik} \rho_{kj}^{(0)} - \rho_{ik}^{(0)} f_{kj}). \tag{3}$$

Here, a phenomenological parameter γ is introduced to simulate the dephasing process.

The computational time for solving Eq. (3) scales as $O(N^{3-4})$ and thus the conventional TDHF is limited to systems of moderate size. However, carbon nanotubes may contain more than thousands of atoms. The LDM method is of $O(N)$ scaling, has been applied to systems with ten thousand atoms, and is thus suitable to calculate the excited state properties of carbon nanotubes. The following approximations

    (a) $\rho_{ij}^{(0)} = 0$ if $r_{ij} > l_0$,

    (b) $\delta\rho_{ij}^{(1)} = 0$ if $r_{ij} > l_1$

are the immediate consequence of the nearsightedness of reduced density matrix, and lead directly to

    (c) $h_{ij}^{(0)} = 0$ if $r_{ij} > l_0$,

    (d) $\delta h_{ij}^{(1)} = 0$ if $r_{ij} > l_1$.

With (a), (b), (c), and (d), the range of the summation $\sum_{k}$ is limited to a finite region for each term of the r.h.s. of Eq. (3). Thus, the computational cost for each $\delta\rho_{ij}^{(1)}$ is finite (i.e. not depending on N). Since only $O(N)$ number of $\delta\rho_{ij}^{(1)}$ are to be determined [because of (b)], the total number of computational steps are then $O(N)$ as well.

IV.   RESULTS

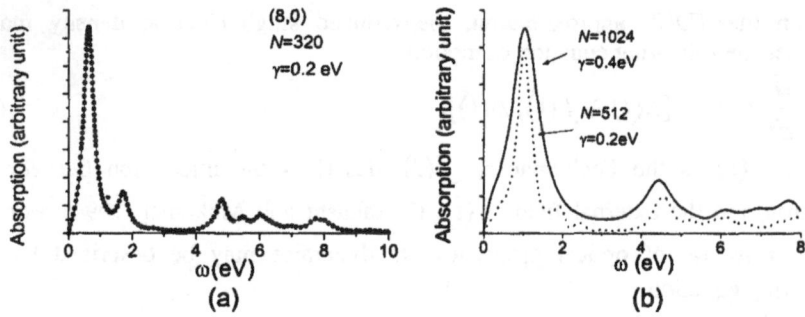

Figure 1: (a)Comparison between TDHF (circle) and LDM (solid line) for N=320. (b) N=512 (dashed line) and N=1024 (solid line).

Figure 1 shows the comparison calculated by full TDHF (circle) and LDM (solid line) for $C_N H_{16}$ (N=320) with chiral vector (8,0). Cut-off lengths are chosen to cover 15 rings for both l_0 and l_1. From the figure, we can clearly see the result with the linear-scaling LDM method agrees to the full TDHF result very well. Thus we employ these cut-off lengths to calculate N=512 (32 rings) and N=1024 (64 rings). The results show the oscillator strength of the first peak becomes less and tends to zero as N increases. We also investigate the tubles (9,0) $C_{576}H_{18}$ and (10,0) $C_{640} H_{20}$. Figure 2 shows absorption spectra for C_{108}, C_{60}, and $C_{48}H_{16}$ based on the PM3 model. There is a marked difference between three spectra. The difference between the first peak position of $C_{48}H_{16}$ and C_{108} indicate that the caps are playing an important roll for the smaller size carbon nanotube. The character of those peaks may be revealed by examining their density matrices.

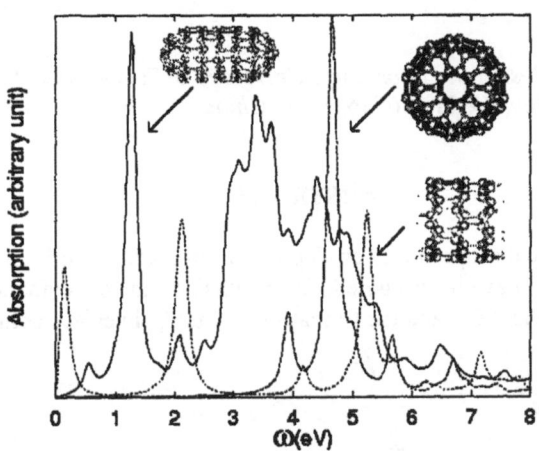

Figure2: Absorption spectra with PM3 model. C_{108} (solid line), C_{60} (long dashed line), and $C_{48}H_{16}$ (short dashed line).

Support from the Hong Kong Research Grant Council (RGC) and the Committee for Research and Conference Grants (CRCG) of the University of Hong Kong is gratefully acknowledged.

Iijima, S., 1991, Nature **354**, pp. 56-58.
Liang, W.Z., Yokojima, S. and Chen, G.H., 1999, Journal of Chemical Physics, **100**, pp. 1844-1855.
Ring, P. and Schuck, P., 1980, *The Nuclear Many-Body Problem* (New York, Springer).
Stewart, J.J.P., 1989, Journal of Computational Chemistry, **10**, pp. 209; 221
Takahashi, A. and Mukamel, S., 1994, Journal of Chemical Physics, **100**, pp.2366.
Yokojima, S. and Chen, G.H., 1998, Chemical Physics Letter. **292**, pp. 379-383.
Yokojima, S. and Chen, G.H., 1999, Physical Review B, in press.
Yokojima, S., Zhou, D.H. and Chen, G.H., 1999, Chemical Physics Letter, in press.

17 Cu-C$_{60}$ Interaction and Nano-composite Structures

J. G. Hou
Structure Research Laboratory, University of Science and Technology of China, Anhui, Hefei 230026, P. R. China

ABSTRACT

Structural and physical properties of co-deposited Cu-C$_{60}$ films with different Cu/C$_{60}$ ratios and grain sizes have been studied. Our results provide the evidences of charge transfer between Cu and C$_{60}$ and the formation of Cu-C$_{60}$ interface compound.

INTRODUCTION

Since the discovery of a method for synthesis of C$_{60}$ in macroscopic quantities (Krätschmer *et al*, 1990), C$_{60}$-metal interaction has been one of the main subjects for in fullerene researches. The interactions of fullerene with metals can be classified according to whether they form compounds or form phase separated solids (Weaver and Poirier, 1994). In the alkali- or alkali-earth metal fullerides, metal atoms are intercalated into the host C$_{60}$ lattice and act as the electron donors, superconductivity was discovered in the intercalated compounds where three electrons are transferred from the metals to each C$_{60}$ molecule (Hebard *et al*, 1991). While for the phase-separated solids, no solid solution is formed due to the high cohesive energy of the metals other than alkali or alkali-earth metals. However, limited intermixing and charge transfer between most of the metals and C$_{60}$ are still possible since the low work functions for most of the metals (Hunt *et al*, 1995, Maruyama *et al*, 1994, Owens *et al*, 1994, Maxwell *et al*, 1995 and Hebard *et al*, 1994). The charge transfer across the metal-C$_{60}$ interface may provide an opportunity to form novel fullerene interface compounds in some kind of low dimensional structures. Motivated by interest in metal-C$_{60}$ compounds, we explored the possibility for making Cu-C$_{60}$ composites via thin film route and studied the structural and physical properties with different techniques .

Experiment

The Cu/C$_{60}$ nano-structure films were grown by co-deposition method. High purity metal and C$_{60}$ (>99.9%) powder were used. By adjusting the source temperatures of metal and fullerene respectively, Cu/C_{60} composite films with different ratio of metal to C_{60} were obtained. The substrates were freshly cut (001) NaCl single crystals, and the temperatures of the substrates were kept at room temperature or 160°C during the film growth. Transmission electron microscope (TEM) was used for the structure characterization. After deposition the substrate was dissolved in distilled water and the film was put on copper grids for structural analysis. DC electrical conductivity of the films was measured by the standard four-probe technique.

Results and discussions

Phase separated composite films were obtained for Cu co-deposited with C_{60}. TEM morphology images and electron diffraction patterns showed that metal nano-crystallites were well dispersed on the C_{60} polycrystalline matrix. However, the volume fraction of the metals as well as the grain size of the metal particles can be controlled by adjusting the deposition rates of C_{60} and Cu. Raman-back scattering spectra of Cu/C_{60} as-grown films have been measured to see if any charge transfer from the metal to C_{60} occurs. The typical spectra of Cu/C_{60} films together with a pure C_{60} film are shown in figure 1. The measurements were performed at 150 K and the laser power was below 50 mW/cm^2 in order to minimise the noise and the effect of photo-polymerization.

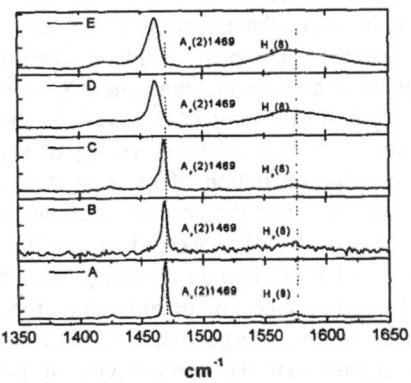

Fig. 1 Raman spectra of C_{60}-Cu co-deposited films

	A	B	C	D	E
$N_{Cu}:N_{C60}$	Pure C_{60}	6	7.5	10	30

Figure 1 shows the as-measured high frequency Raman spectra. In comparison with pure C_{60} film, the Raman spectra of the Cu/C_{60} films are changed considerably due to the Cu/C_{60} interactions. Obvious difference are that the peak width of the pentagon pinch mode ($A_g(2)$) becomes wider and the peak position shifts to the lower frequency as the atomic ratio of Cu/C_{60} increases. The vibrational frequencies of the softened modes are about 8-10cm^{-1} lower than that of the pristine mode. If the softening of $A_g(2)$ modes could be attributed to the charge transfer, we can infer that ~1.5 electrons are transferred to per C_{60} molecule from metal using the calibration of ~6cm^{-1} shift per electron transfer (David *et al*, 1991).

Electrical conductivity of the films was measured by the standard dc four-probe technique. Fig.2a shows the typical $\ln(\sigma/\sigma_{RT}) \propto (1/T)$ curves of Cu-C$_{60}$ films prepared at 160 °C with different Cu/C$_{60}$ ratio. Semiconducting behaviours of conductivity versus temperature were typically observed for most of the films in which the metal volume fraction is under the percolation threshold except the film of high Cu/C$_{60}$ ratio(N_{Cu}/N_{C60}=180). There is a transition at about 250 K for all the semiconducting films due to the phase transition of solid C$_{60}$(Gugenberger *et al*, 1992). It is noted that the linear relationships of $\ln(\sigma/\sigma_{RT}) \propto (1/T)$ were observed when the temperatures are lower than 250K, and $\ln(\sigma/\sigma_{RT}) \propto (1/T^{1/4})$ relations were observed when temperatures higher than 260K(Fig. 2b). The magnitude of the conductivity increases, while the activation energy decreases with the increasing of Cu ratio. For the films prepared at room temperature, $\ln(\sigma/\sigma_{RT}) \propto T^{-1}$ relation has been observed above and below the transition temperatures. The values of conductivity of these films (typically in the range of 10^{-2} to 10^{2} S-cm^{-1}) are much higher than that of the pristine C$_{60}$ film solid(~10^{-7} S-cm^{-1}).

The relation of $\ln(\sigma) \propto T^{-1/2}$ has been commonly observed for metal-insulator composite films independent of the kinds of metals and insulators (Abeles *et al*, 1975). So the transport measurement results indicate that the electrical transport of the Cu/C$_{60}$ does not dominate by the tunnelling between the isolated Cu particles. The transport mechanism of these composite films is likely the hopping conductivity which typically observed in amorphous semiconductors since the activation energies are very small, typically in the range of 50 to 1 meV. There are two possible hopping conduction mechanisms if states at the Fermi energy of a condensed electron gas are localized (Mott, 1990): the excitation of the carrier to a mobility edge and the thermally activated hopping. For the former mechanism, $\sigma = \sigma_0 \exp(-(E_c - E_f)/k_B T)$. While the relation of $\ln(\sigma) \propto T^{-1/4}$ is observed for the thermally activated hopping. According to the results of TEM, Raman scattering and electrical transport, we may inference that metal-C$_{60}$ interface compounds were formed in the Cu/C$_{60}$ nano-composited films. Due to the charge transfer, the conductivity of the interface compound is much higher that the conductivity of pristine C$_{60}$ solid and the tunnelling conductivity of discontinuous granular metals films with the same metal volume fraction (Abeles *et al*, 1975). The structure of the possible interface compound seems depend on the substrate temperature. For the films prepared at high temperature, the relation of $\ln(\sigma) \propto T^{-1/4}$ was observed when temperatures are higher than 250 K. The reason that why the phase transition of C$_{60}$ leads to the different transport mechanism is not clear yet. It is also noted that the relation of $\ln(\sigma) \propto T^{-1/4}$ is usually observed when the temperature is very low so that the density of the state (N(E$_f$)) is finite but states are localized near the Fermi energy. So what is the relation of rotation of C$_{60}$ to the localization of states in the Cu/C$_{60}$ films is a very interesting question and further studies are underway.

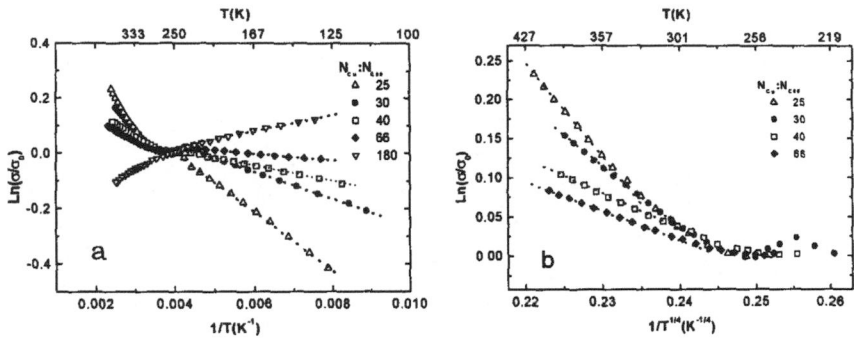

Fig 2 Temperature dependent normalized conductivities of the films.

CONCLUSIONS

Interesting structure and physical properties were observed in Cu/C_{60} nano-structures The strong interaction between Cu and C_{60} not only leads to the charges transfers from metal to C_{60} at the interface, but also form some Cu/C_{60} interface compound which exhibits hopping conductivity.

REFERENCE

Abeles, B., Ping Sheng, Coutts, M.D., and Arie, Y., 1975. *Advanced Physics*, **24**, pp. 497.

David, W.I.F., Ibberson, R.M., Matthewman, J.C., Prassides, K., Dennis, T., Hare, J.P., Kroto, H.W., Taylor, R., and Walton, D., 1991. *Nature*, **353**, pp. 147.

Gugenberger, F., *et al*, 1992. *Physcal Review Letters*, **69**, pp. 3774.

Hebard, A.F., Ruel, R.R., and Eom, C.B., 1996. *Physcal Review*, **B54**, pp. 14052.

Hebard, A.F., Rosseinsky, M.J., Haddon, R.C., Murphy, D.W., Glarum, S.H., Palstra, T.T.M., Ramirez, A.P., Kortan, A.R., 1991. *Nature*, **350**, pp. 600.

Hunt, M.R.C., Modesti, S., Rudoli, P., Palmer, R.E., 1995. *Physcal Review*, **B51**, pp. 10039.

Krätschmer, W., Lamb, L.B., Forstiropoulos, K., Huffman, D.R., 1990. *Nature*, **347**, pp. 354.

Maruyama, Y., Ohno, K., Kawazoe, Y., 1994. *Physcal Review*, **B52**, pp. 2070.

Maxwell, A.J., BrÜhwiler, P.A., Andersson, S., Arvanitis, D., Hernnäs, B., Karis, O., Mancini, D.C., Martensson, N., Gray, S.M., Johansson, M.K., Johansson, L.S.O., 1995. *Physcal Review*, **B52**, pp. R5546.

Mott, N.F., 1990. *Metal-insulator transitions* (Taylor & Francis Ltd.).

Weaver, J.H., Poirier, D.M., 1994, Ch.1, Fullerene Fundamentals. In *Solid State Physics*, Vol.48, edited by Ehreneich, H. and Spaepen, F.(Academic, Cambridge).

CONCLUSIONS

REFERENCES

Part 3

FUNCTIONAL
NANOSTRUCTURED
MATERIALS

18 Microstructured Particles for Electromagneto-rheological (EMR) Applications

W. Y. Tam, W. Wen, N. Wang, G. H. Yi, M. M. T. Loy, H. Ma,
Z. Lin, C. T. Chan and P. Sheng
*Department of Physics, Hong Kong University of Science and
Technology, Clear Water Bay, Kowloon, Hong Kong*

1.1 ABSTRACT

Common electrorheological (ER) fluids are suspensions of simple dielectric particles in non-conducting fluid where the fluid properties are controllable through the application of an external electric field. Guided by first-principles calculations based on the minimization of free energy density, we have fabricated novel doubly-coated micro-particles with a conducting inner layer and an insulating outer layer for ER applications. In good agreements with the calculations, orders of magnitude enhancement in the yield stress under an electric field was obtained using the doubly-coated particles when compared to that of bare or singly-coated (without the conducting layer) particles. We have also fabricated multiply-coated particles of glass spheres that exhibit simultaneous electro- and magneto- responses. Under crossed electric and magnetic fields, a transition from the body-centered-tetragonal (BCT) structure to the face-centered-cubic (FCC) structure was observed when the ratio between the magnetic and the electric fields exceeded a minimum value. Both the FCC crystal formation and the BCT-FCC transition scenario are correctly predicted by the first-principles calculations.

1.2 INTRODUCTION

Electrorheological (ER) fluids are a class of materials whose rheological properties are controllable through the application of an electric field (Havelka and Filisko, 1995). The most common type of ER fluids is a colloid of solid dielectric particles dispersed in a fluid. Here the change in the rheological properties is known to be accompanied by a structural transformation whereby the initially dispersed solid particles are aggregated into dense columns spanning the sample and aligned along the direction of the applied electric field (Tao and Sun, 1991a and 1991b). These dense columns, with body-centered-tetragonal (BCT) meso-crystallites, are the source of ER fluids' increased viscosity when shearing occurs perpendicular to the electric field.

Electrorheological fluids are known to hold great potential for many applications such as in shock-absorbing vibration dampers, clutches, and robotics (Goldstein, 1990; Korane, 1991). However, large scale applications of ER fluids have been slow due to the lack of ER systems with the required yield stress, stability of the ER effect over extended period, ER dynamic range, and acceptable wear resistance, among other conditions. Moreover, it is difficult to find a single material that can satisfy the above requirements. Artificially structured particle is a solution to solve some of the problems.

In this paper, we present first principles calculations for simple dielectric particles ER fluids and apply the calculations to a novel ER system consisting of doubly-coated particles in a dielectric fluid. The doubly-coated particles consist of a core, a metal inner layer, and a dielectric outer layer. Such particles offer a much broader range of material and controllable parameter options to both enhance the ER effect and fulfill many of the application constraints. In particular, the choice of the core material, which controls the density of the particle, can directly impact the sedimentation problem. The metallic coating would enhance the ER effect as shown below. The right choice of the outer dielectric layer material would serve to electrically insulate the particle, provide mechanical wear protection, as well as further enhance the ER effect through its high dielectric constant. We have successfully fabricated such doubly-coated particles and demonstrated the expected enhancement of the ER effect (Tam *et al.*, 1997). In addition, we have also extended our techniques to fabricate multiply-coated glass spheres that exhibit appreciable electro- as well as magneto-rheological responses (Wen *et al.*, 1998). Under crossed electric and magnetic fields, a transition from the BCT structure to the face-centered-cubic (FCC) structure was observed when the ratio between the magnetic and the electric fields exceeded a minimum value. Both the FCC crystal formation and the BCT-FCC transition scenario are correctly predicted by the first principles calculations of minimizing the combined electrostatic and magnetostatic free energy. In section 1.3 we present the first principles calculations for simple dielectric particles and followed by the application to the doubly- and multiply-coated particles. Section 1.4 will describe the fabrication of the doubly- and multiply- coated particles, the experimental measurements of the shear stress and the dielectric constants, and the comparisons with the calculations. Conclusions will be presented in section 1.5.

1.3 FIRST PRINCIPLES APPROACH AND FORMALISM

The structural transformation of dispersed solid particles to dense columns can be understood as follows. Due to the difference in the dielectric constants of the solid particles and the liquid, each particle is polarized in the presence of an applied electric field. The particles then interact, to a good approximation, with dipole-dipole interaction. To lower the interaction energy the particles arrange in periodic structure, i.e., a solid. It has been shown that, based on the dipole-dipole interaction, the lowest-energy periodic structure is the body-centered tetragonal (BCT) structure (Tao and Sun, 1991a). This result has later been verified by more accurate calculations (Clercx and Bossis, 1993; Friedberg and Yu, 1992; Davis, 1992). To put our understanding of the ER fluids system on a more rigorous basis, it is desirable to carry out first principles calculations (Ma, *et al.* 1996). The basis of such a theoretical framework is the observation that the operating (applied electric field) frequencies of the ER fluids are generally below 10^5 Hz, and the dielectric particle size is below 100 μm. That means comparison between the electromagnetic wavelength (on the order of kms) and the scale of inhomogeneities in an ER fluid would firmly place the system in the quasistatic limit, where the electrical response of the system is captured by an effective complex dielectric constant tensor, $\bar{\varepsilon}$. The electrostatic Gibbs free energy of

the system is then given by $F_E = -\mathrm{Re}(\bar{\varepsilon}_{zz} E^2 / 8\pi)$, where $\bar{\varepsilon}_{zz}$ is the zz component of the effective dielectric tensor, z being the applied electric E field direction. Thus as far as the electrostatic free energy is concerned, the equilibrium state is one where $\mathrm{Re}(\bar{\varepsilon}_{zz})$ is maximized with respect to the positions of the solid particles. We denote that state the electrostatic ground state of the ER fluid. The ER mechanical properties, such as the shear modulus and the static yield stress in the high field regime, may be obtained by perturbations away from the electrostatic ground state.

For the calculation of $\bar{\varepsilon}$, we use the Bergman-Milton representation (Bergman, 1978 and 1979; Milton, 1980 and 1981). For simplicity, we first consider a two-component system consisting of spherical solid particles of radius R and complex dielectric constant ε_1 dispersed in a fluid characterized by ε_2. The case of coated particles, to be described later, represents a generalization of the two-component system formalism. The electrostatic problem to be solved is given by

$$\nabla \cdot \varepsilon(\bar{r})\nabla \phi = 0, \tag{1}$$

where $\phi(\bar{r})$ is the electrostatic potential. By expressing $\varepsilon(\bar{r}) = \varepsilon_1 \eta(\bar{r}) + \varepsilon_2 [1 - \eta(\bar{r})] = (\varepsilon_1 - \varepsilon_2)\eta(\bar{r}) + \varepsilon_2$, where $\eta(\bar{r})$ is the characteristic function for the solid component, defined as having the value 1 at those spatial points occupied by the solid particles, and zero otherwise, a formally *exact* solution to Eq. (1), given the condition of $\Delta\phi / \ell = E = 1$ in the z direction, may be written as

$$\phi = -(1 - \frac{1}{s}\Gamma)^{-1}z = -s\frac{z}{s - \Gamma} \quad , \tag{2}$$

where $s = \varepsilon_2 / (\varepsilon_2 - \varepsilon_1)$ is the only relevant material parameter in the problem, and $\Gamma = \int d\vec{r}'\eta(\vec{r}')\nabla'G_o(\vec{r} - \vec{r}') \cdot \nabla'$ is an integral-differential operator, with $G_o(\vec{r} - \vec{r}') = 1/4\pi |\vec{r} - \vec{r}'|$ denoting the Green's function for the Laplace equation.

By defining the inner product operation as $\langle \phi | \psi \rangle = \frac{1}{V}\int d\vec{r}'\eta(\vec{r}')\nabla'\phi^* \cdot \nabla'\psi$, where V denotes the sample volume, it becomes possible to write the effective dielectric constant in the Bergman-Milton representation (Bergman, 1978 and 1979; Milton, 1980 and 1981):

$$\frac{\bar{\varepsilon}_{zz}}{\varepsilon_2} = 1 - \frac{1}{V}\sum_n \frac{|\langle z | \phi_n \rangle|^2}{s - s_n} = 1 - \sum_n \frac{f_n^z}{s - s_n} \quad . \tag{3}$$

Here s_n and ϕ_n are the nth eigenvalue and eigenfunction of the operator Γ.

The remarkable feature about the representation, Eq. (3), is that the geometric information is separated from the material information, in contrast to approaches that involve the direct numerical solution of the Laplace equation. This separation means that as far as the dielectric properties are concerned, whereas the material information appear only in s, the microstructural information are given by the $|\langle z | \phi_n \rangle|^2 / V = f_n^z$ and the locations of the poles s_n, both of which are known to be real. Furthermore, it has been proved that s_n must lie in the interval [0,1] (Bergman, 1978 and 1979; Milton, 1980 and 1981). Once the geometric information is obtained, it becomes simple to calculate the effective properties that can be derived from energy considerations, *including those due to the conductivities and their associated frequency dependencies*, as these factors appear only in s. If either ε_1 or ε_2 is complex and frequency dependent, then the resulting s, and consequently $\bar{\varepsilon}_{zz}$, will be complex and frequency dependent. Whereas the imaginary part of $\bar{\varepsilon}_{zz}$ characterizes the overall electrical dissipation of the system, the frequency dependence of the real part of $\bar{\varepsilon}_{zz}$ is what gives rise to the frequency dependencies of the yield stress and the shear modulus (Ma, et al., 1996). It should be emphasized that the present formulation is rigorous and *includes all the multipole interactions and (self-consistent) local field effects*. We have performed numerical calculations of $\bar{\varepsilon}_{zz}$ with high precision for six periodic structures--the body-centered tetragonal (BCT), the face-centered cubic (FCC), the hexagonal, the body-centered cubic, the simple cubic, and the diamond. It was found that at any given concentration of the particles, body-centered tetragonal has the largest $\bar{\varepsilon}_{zz}$ and face-centered-cubic is a close second, with the rest decreasing in the order given above. This fact is in agreement with prior calculations based on the dipole interactions (Tao, 1991a and 1991b; Clercx and Bossis, 1993; Friedberg and Yu, 1992; Davis, 1992).

An important point to be emphasized is that the lattice structure derived from the ER effect is stabilized only by the hard core repulsion between the solid spheres. That is, even at equilibrium there is an electrostatic contraction pressure on the particles. As a result, the ER lattice structure has no bulk modulus in the usual sense because the first derivative of the electrostatic potential (as seen by the particles) does not vanish along those distortion paths that would pull the spheres apart. However, for those distortion paths that maintain the contacts between any given spheres and its neighbors (such as the shear distortion defined below) the definition of a modulus becomes possible.

In order to calculate the shear modulus and the yield stress, it is necessary to perturb the system away from its lowest electrostatic ground state. For the BCT structure, shearing in a direction perpendicular to the z-axis means not only a tilt of the c-axis away from the electric-field direction by an angle θ, but also a distortion[1] in the lattice constants c and a given by $c/R = 2/\cos\theta$, $a/R = [8-(c^2/2R^2)]^{1/2}$. Numerical evaluation of the stress-strain relation, i.e., $\mathrm{Re}\{|\varepsilon_2|^{-1}(\partial\bar{\varepsilon}_{zz}(\theta)/\partial\theta)\}$ versus θ shows a maximum which corresponds to the static yield stress. Further increase in the strain would make the stress decrease, i.e., the structure becomes unstable. The slope of the linear region in the small θ limit gives the shear modulus. In actual experiments, it is often the case that columns are formed. However, the difference in the free energies of the columns state and the true ground state is small (at most 0.1% of the total electrostatic energy) (Ma, *et al.*, 1996). As a result, it is expected that the calculated shear modulus or yield stress should be relatively robust and accurate for the experimentally encountered columnar states. Moreover, it is now possible to derive the upper bounds on the effective dielectric constant, the shear modulus, and the static yield stress using the above formalism as reported in Ma *et al.* (1996).

1.3.1 Doubly-coated Particles

For the doubly-coated particles system, the above formalism is generalized as follows. First, the dielectric constant is now given by $\varepsilon(\vec{r})/\varepsilon_\ell = 1 - \sum_i^3 \eta_i(\vec{r})/s_i$, where $s_i = \varepsilon_\ell/(\varepsilon_i - \varepsilon_l)$, ε_ℓ denotes the liquid dielectric constant (replaces ε_2

---

[1] Here the distortion is treated as symmetrical in the x and y directions since that is lower in free energy than any distortions that are anisotropic in x and y.

in the two-component case), ε_i is the dielectric constant of the ith solid materials ($i = 1,2,3$), and $\eta_i(\bar{r})$ is the characteristic function of solid material component i, taking the value 1 in the region of component i, and zero otherwise. With the same imposed voltage condition as before, the effective dielectric constant is given by

$$\frac{\bar{\varepsilon}}{\varepsilon_\ell} = 1 - \frac{1}{V}\sum_{i=1}^{3}\frac{1}{s_i}\int d\bar{r}\frac{\partial\phi(\bar{r})}{\partial z}\eta_i(\bar{r}) \, , \tag{4}$$

where $|\phi\rangle = [1 - \sum_i \Gamma_i / s_i]^{-1}|z\rangle$ is the formal solution to Eq. (1), with $\Gamma_i = \int d\bar{r}'\eta_i(\bar{r}')\nabla'G(\bar{r},\bar{r}')\cdot\nabla'$. The electrostatic ground state for the coated particles remains BCT, just as in the two-component case. The calculation of either the shear modulus or the static yield stress follows the same procedure as before.

1.3.2 Multiply-coated EMR Particles

In our calculations for the multiply-coated EMR particles, we assume that the spheres form columns with diameter about 10 times the diameter of the spheres, and the preferred arrangement of the spheres inside the columns is determined through energy minimization. We consider the combined Gibbs free energy densities of the EMR fluid, $F = F_E + F_H$, where F_E and F_H are the electrostatic and magnetostatic parts of the free energy, respectively. F is a function of the external fields, the arrangement of the spheres, as well as the dielectric and magnetic properties of the coated spheres and the fluid medium. The electrostatic free energy density is again given by $F_E = -\mathrm{Re}(\bar{\varepsilon}_{zz}E^2/8\pi)$. Here, $\bar{\varepsilon}_{zz}$ is the zz component of the effective dielectric tensor for the composite system. For the magnetic part, each EMR sphere is assumed to have a permanent moment $\bar{m}_i$.

The magnitude of this moment is assumed to be fixed, but its orientation can change depending on the arrangement of the spheres and the strength of the H field, as determined by the minimization of the magnetostatic free energy density:

$$F_H = F_m - \vec{M}\bullet\vec{H} + 2\pi\eta_{\alpha\beta}M_\alpha M_\beta \, . \tag{5}$$

The first term, $F_m = -\dfrac{1}{2\Omega}\sum_i \bar{m}_i \bullet \sum_{j \neq i}\dfrac{1}{r_{ij}^3}\Big[3\hat{n}(\hat{n}\bullet\bar{m}_j) - \bar{m}_j\Big]$, is the dipolar interaction due to the permanent moments. The second term is the interaction between the applied H-field and the magnetization $\vec{M}$ of the entire EMR system.

The third term gives the depolarization effect, where $\eta_{\alpha\beta}$ are the demagnetization factors of the rectangular EMR cell (taken to be diagonal and $\eta_{xx} = \eta_{yy} = \frac{1}{4}$ and $\eta_{zz} = \frac{1}{2}$). To evaluate the F_H, we determine the orientation of each individual $\vec{m}_i$ through a spin dynamics simulation. A set of dynamical equations is derived from a Lagrangian $L = T - F_H$, where T is an auxiliary kinetic energy that is dissipated slowly to allow the spins to evolve towards an energy minimum. We assume the cylindrical columns to be arranged in a square lattice. Each spin is allowed to vary freely, with the periodic boundary condition imposed in the z-direction. Our approach is noted to differ from the previous theoretical considerations (Bossis *et al.*, 1994; Tao, 1998) in the following aspects: (i) the macroscopic columnar structure is explicitly considered; (ii) the dielectric part of the interaction energy is computed with the Bergman-Milton formalism rather than the usual dipole approximation; (iii) permanent (rather than induced) magnetic moment is considered in better correspondence with the material properties; (iv) we do not assume uniform alignment of the magnetic moments, which is unphysical in our case. The BCT, the FCC and the intermediate structures inside the mesocrystallite columns can all be described by a body-centered-orthorhombic unit cell with two spheres per cell. Thus the Gibbs free energy can be determined for the EMR fluid under both electric and magnetic fields.

1.4 FABRICATIONS AND EXPERIMENTAL RESULTS

The fabrications for the doubly- and multiply- coated particles are summarized below and details should be refered to Tam *et al.* (1997) and Wen *et al.* (1998). Experimental results will also be presented and compared to the calculations as stated in section 1.3.

1.4.1 Doubly-coated Particles

We now describe the fabrication process for the dielectric/metal doubly-coated particles (DMP) (Tam *et al.*, 1997). The core particles are solid glass spheres commercially available in various sizes. We used spheres with diameters 1.5 (± 0.1) μm and 50 (± 7.5) μm. Conventional electroless plating process was used to deposit the inner conducting nickel layer (Severin *et al.*, 1983). The outer insulating layer material used was titanium oxide, chosen for its hardness, resistance to wear, and high dielectric constant. This outer layer was deposited using a sol-gel process (Briker and Scherer, 1990). It is crucial to avoid particle coagulation during the process, and this was achieved by adding excess sugar solution to form a buffer, which could subsequently be removed by heating. The sugar solution, in addition, formed a foam structure during heating which further separated the particles. After heating ($500°C$ for 10 hours) to remove all organic components, the DMP's were collected as ashes and dispersed in silicon oil to form the DMP ER fluid. The heating also serves to properly anneal the TiO$_2$ coating, with the very desirable properties of high dielectric constant, excellent adhesion and hardness.

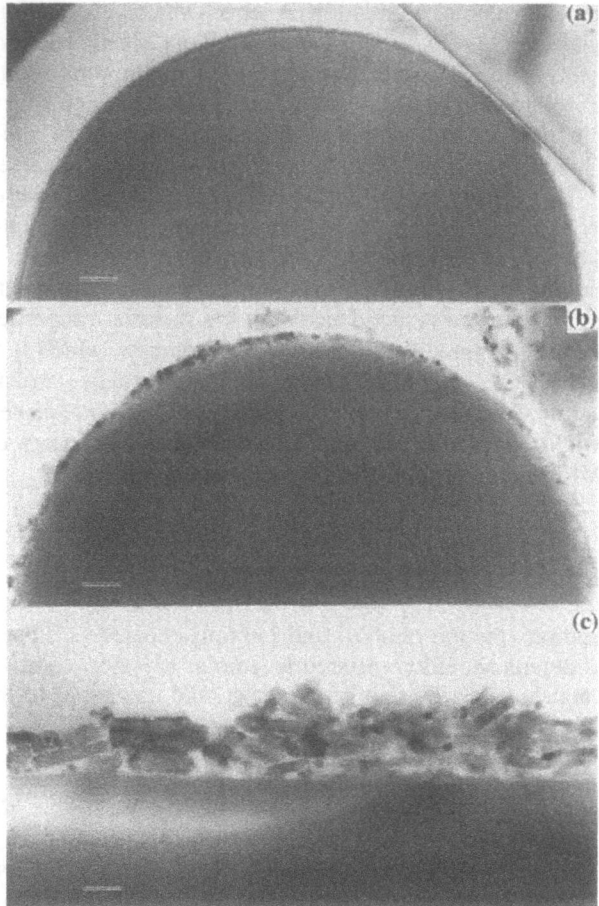

Figure 1. Cross sectional electron micrographs for (a) 1.5 μm nickel coated, (b) 1.5 μm doubly-coated, and (c) 50 μm doubly-coated particles. The scale bar is 100 *nm* and 25 *nm* for (a)/(b) and (c), respectively. While in (a), the metal/glass interface is clearly visible, diffusion of Ni atoms, probably from the heating process following the sol-gel process, blurred this boundary, as shown in (b) and (c). The TiO$_2$ coating thickness is seen to be in the range of 10 *nm* to 30 *nm* for (b) and 25 *nm* to 60 *nm* for (c). Note also the darker spots seen only in the TiO$_2$ coatings in (b) and (c).This leads us to believe that the TiO$_2$ coating was at least partially crystalline, the dark spots being regions with high scattering efficiency when the Bragg condition was satisfied when the electron micrograph was taken. With higher magnification, the apparently microcrystalline morphology of the TiO$_2$ layer can be seen even more clearly in (c).

A 1.5 μm particle with only the nickel coating is shown in Figure 1(a). It is seen that the metallic coating was on the order of 100 $\overset{o}{A}$ in thickness with excellent uniformity. Figure 2(b) shows a doubly-coated particle at the completion of the process (after heating.) In contrast to Figure 1(a), the TiO_2 coating has a much greater variation in thickness, ranging from 100 $\overset{o}{A}$ to 250 $\overset{o}{A}$ in this case. Figure 1(c) shows, at higher magnifications, a 50 μm doubly-coated particle showing the annealed crystalline structure of the TiO_2 outer coating. Comparison between Figures 1(a) and 1(b) or 1(c) also indicated the disappearance of the metal/glass boundary clearly visible in Figure 1(a). This is most likely due to diffusion of the nickel atoms during the heating process after the TiO_2 coating process. However, as we will see in yield stress measurements below, this conducting layer, while not visible in the picture, is sufficiently continuous to fulfill its mission of expelling the electric field (from the interior of the particles) and led to behavior markedly different from particles without such an inner conducting coating. As a further check, we have performed compositional analysis of the particles. The results are in general agreement with the above picture.

To study the ER effect we used a parallel plate viscometer to measure the yield stress of the DMP/silicon oil ER system. Water was removed from the system by a 24 hours heating process at 120^0 C prior to each measurement. The yield stress was measured with a 50 Hz AC field applied between the parallel plates, and using samples with particle/fluid fraction fixed to 0.2. Figures 2(a) and 2(b) show the measured static yield stress for a DMP/silicon oil ER system using the 1.5 μm and 50 μm diameter particles, respectively, as a function of electric field strength. Static yield stress over 2000 Pa was obtained at 2 kV/mm for the DMP system. As comparisons, we also showed the static yield stress measured for the bare glass particles as well as glass particles coated only with TiO_2 without inner metallic coatings (see also Figure 2(c) with expanded scale for Figure 2(a)). The ER enhancement of the DMP system was over two orders of magnitude. It is seen that the DMP system followed generally the expected electric field squared dependence. These DMP's were found to be extremely robust and can sustain both the high electric field and high stress which are prerequisites for any application. The size dependence is seen to be in general agreement with the intuitive picture presented above. In Figures 2(a) and 2(b) are also plotted theoretical predictions calculated from TiO_2 and silicon oil dielectric constant values of 85 and 2.5, respectively, with the thickness of the TiO_2 coating thickness indicated besides each of the theory curves (solid lines). The metal dielectric constant is taken to be infinite, which also makes its thickness immaterial. Good agreements are seen for both particle sizes. Using the same theoretical model, we have also calculated the yield stress for bare glass particles and TiO_2 coated particles (without the inner metallic coating) The results, together with the experimental results for such particles, are shown in Figure 2(c) for the 1.5 μm particles. As a reference, we have also calculated the yield stress expected for TiO_2 particles of the same size, shown as dashed lines in Figures 2(a) and 2(b). Both the ability of the first principles calculations to account for the size and coating thickness dependencies of the static yield stress, and the efficacy of the doubly-coated DMP system, are clear from the figures.

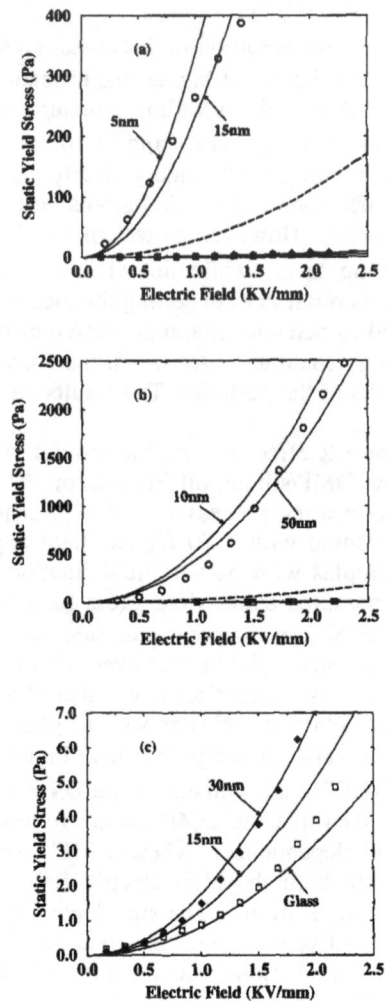

Figure 2 The measured (symbols) and calculated (lines) static yield stress of ER fluid using (a) 1.5 μm doubly-coated particles, (b) 50 μm doubly-coated particles, and (c) expanded scale of (a), uncoated particles (open squares) and singly coated with TiO_2 only (solid squares). As explained in the text, the solid line is yield stress obtained from first principles calculations using known dielectric constants and measured geometric information of the particles. Due to the variation of TiO_2 coating thickness as seen in Figures 1(b) and 1(c), the calculation was done for a range of TiO_2 thicknesses, as labeled in the figure. As a comparison, we have also calculated for the case of solid TiO_2 particles of the same sizes (dashed line) showing that the doubly-coated particles have far superior static yield stress.

1.4.2 Multiply-coated EMR Particles

To fabricate the multiply-coated particles 34 ($\pm$ 2) μm glass spheres were first coated with a layer of approximately 2 μm Ni using method as mentioned above for the doubly-coated particles and then with a layer of PZT using a sol-gel method to prevent Ni oxidation in the subsequent annealing process, necessary for obtaining the desired magnetization (Wen *et al.*, 1998). The PZT-coated spheres were heated in vacuum at 400° *C* for 2 hours and then annealed at 550° *C* for three hours. While not absolutely necessary, we have further coated another layer of Ni, followed by a layer of TiO_2, with processing steps as used in the doubly-coated particles stated above. The final size of the multiply-coated glass was 45 ($\pm$ 4) μm. The cross sectional scanning electron microscope (SEM) pictures of the EMR spheres, together with the response to a small magnet, are shown in Figure 3.

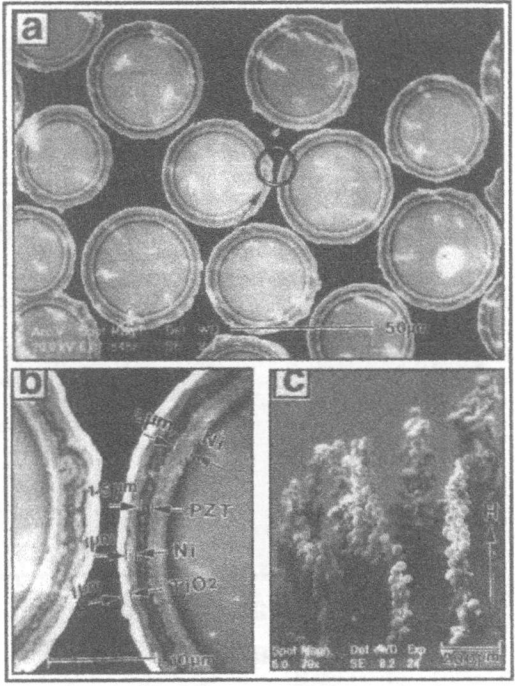

Figure 3. (a) Cross sectional SEM picture of the coated spheres. The apparent size variation is caused by deviation of the spheres' centers from the cutting plane. The arrow points to a circular region detailed in (b); which shows detailed thickness of the four coatings. From the inside out: 2 μm Ni, 1.5 μm PZT, 1 μm Ni, and finally 1 μm TiO_2. (c) The coated EMR spheres under the influence of a small magnet.

Measurements of dielectric constant of EMR fluids were carried out by mixing the EMR particles with silicon oil, and placing the sample in a cell with four electrodes. Two electrodes top and bottom, separated 3 *mm* apart, were used for the high E-field and while two side electrodes, separated 6 *mm* apart, were connected to a LCR meter for dielectric constant measurement. The whole cell was placed in the central region of an electromagnet such that the direction of the applied H-field was perpendicular both to the E-field and the direction of the dielectric measurement. Without external fields, the suspension had the appearance of a random dispersion. At an E-field of 1 to 2 *kV/mm*, columns appeared, with diameters typically ranging from 7 to 9 particles. At fixed E-field, the structural changes induced by the H-field inside the meso-crystallites were monitored by measuring the small dielectric constant changes. The results are summarized in Figure 4 for a sample with 20% solid volume fraction, for four values of the E-field.

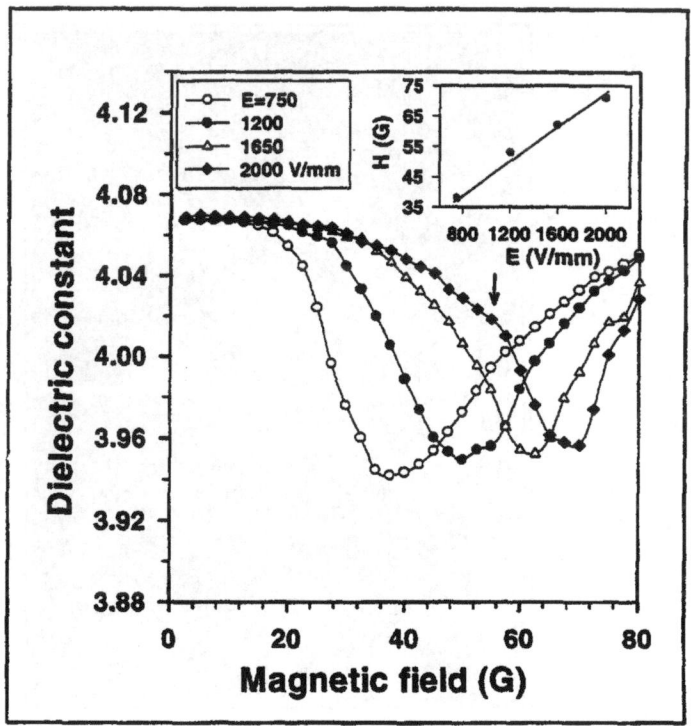

Figure 4. The sample dielectric constant measured along the *y*-direction (perpendicular to both the E- and H-fields) as a function of applied H-field, under four different E-fields. The upper right inset shows that the position of the minimum is a linear function of the applied E-field. The arrow indicates the H-field value where the FCC structure was observed.

They have a characteristic dip on the order of 3%, with the minimum position increasing linearly with increased E-field, as shown in the upper-right inset. The dielectric constant at the small H-field region was flat and reversible.

For the curve with E=2 *kV/mm*, this region occurred for H < 30 *G*. Irreversibility sets in at H-fields greater than this value. We obtained cross-sectional micrographs by freezing in solid epoxy the configurations at various H-field values, and cutting the resulting samples. Four such micrograph pictures are shown in Figure 5, with E=2 *kV/mm*. Figures 5(a) and (b) are for the configuration under zero H-field, cut along the (001) plane (Figure 5(a)) and the (110) plane (Figure 5(b)). Taken together, they give direct evidence to the BCT structure. Figures 5(c), cut along the (011) plane, and 5(d), cut along the (110) plane, are the configuration at H=54 *G*. They clearly indicate a square lattice in the (001) plane (the FCC(110)) and a hexagonal lattice in the (110) plane (the FCC(111)), both are the signatures of a FCC structure. Other cuttings were taken at 20 *G*, 30 *G*, 35 *G*, 38 *G*, and 50 *G*. Together with the dielectric constant measurements, they give the following picture of the BCT-FCC transformation.

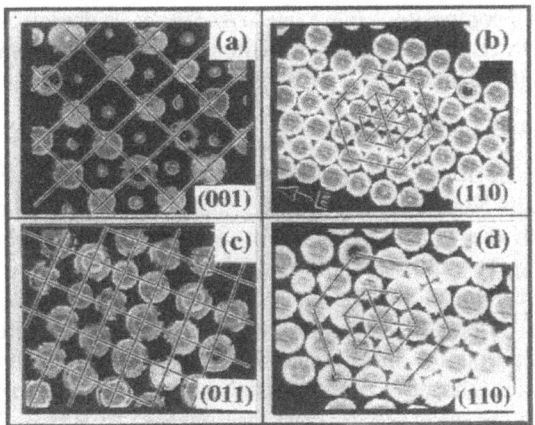

Figure 5. Panels (a) and (b) are SEM cross sectional pictures for a sample frozen (in epoxy) at E=2 *kV/mm* with zero H-field; while (c) and (d) are cross sections for a sample frozen at E=2 *kV/mm* and H = 54 *G*. See text for details.

At H < 30 *G*, only the BCT structure was seen. From 30 *G* to 50 *G*, there was an apparent coexistence of local BCT structures and local non-BCT structures. This coincides approximately with the onset of irreversibility in the dielectric constant measurement. At 54 G only the FCC structure was seen. At H > 54 G, we see visually the columns being pulled apart and forming a "fractal-like" structure at the minimum of the dielectric constant. Even higher H-fields result in the re-formation of the columns aligned in the H-field direction, similar to the observation reported in Koyama (1996). We note here that the cohesive energy per particle (relative to random dispersion) in our system is at least 5 orders of magnitude larger than the thermal energy, and hence temperature effects are negligible. The transition between two structures of distinct symmetries was accomplished by the cooperative displacements of the spheres without any long range diffusion. As such, it can be regarded as a Martensitic transformation. As the BCT and the FCC structures are close in energy under an E-field alone, the structural transition is induced by varying the relative strengths of the two external fields. However, as will be exposed later, the macroscopic columnar structure of these mesocrystallites plays a subtle but important role in deciding the FCC orientation.

To simplify the calculations for the multiply-coated EMR spheres, we approximate the dielectric polarizability of the EMR spheres by that of the TiO_2 coating (which is a lower bound due to the inner Ni coatings). The overall $\bar{\varepsilon}_{zz}$ can be found as $\bar{\varepsilon}_{zz} = f_c \varepsilon_{zz} + (1 - f_c)\varepsilon_2$, where f_c the volume fraction of the mesocrystallite columns, fixed at the experimental value of $0.2/p_c$, with p_c the solid fraction in the mesocrystallites. The $\bar{\varepsilon}_{zz}$ of the mesocrystallite, treated as a two-component composite, may be rigorously calculated for any given microstructure through the Bergman-Milton representation (Bergman, 1978 and 1979; Milton, 1980 and 1981), with the input parameters of $\varepsilon_2 = 2.5$ for the dielectric constant of the fluid and $\varepsilon_1 = 85$ for that of TiO_2.[2] The magnetic moment of the EMR spheres has approximately 1.5×10^{-6} *emu* in magnitude estimated from its Ni coating (assuming a magnetization of 58 emu/g). We align the c axis with the E-field (z-direction) and the a axis with the H-field (x-direction). Using the expressions for the Gibbs free energy densities for the EMR spheres as stated in section 1.4.2, we found that up to H = 60 G, the minimum energy state is associated with $c = 2R$ (i.e. touching spheres in the E-field direction). In addition, the spheres must be in physical contact with each other, as required by the fact that the compressional electrostatic and magnetostatic forces are stabilized by the steric repulsion of the spheres.[3] At H = *0 G*, the ground state

---

[2] The dielectric constant of TiO_2, depending on its crystalline orientation with respect to the applied E-field, is 85 or 170. Here, for lack of information supporting a highly crystalline state, we use the lower value.

[3] If we allow the spheres to come apart, there are multiple deformation paths that can bring the columnar mesocrystallites into other arrangements that can have lower energy than the FCC structure (such as very oblate columns with HCP internal structure). However, the transformation barrier involved is orders of magnitude higher than the thermal energy, and such paths have no physical relevance.

is the BCT structure, with $a/c = \sqrt{6}/2$ (and $b=a$), as seen experimentally and in agreement with prior works (Tao and Sun, 1991a; Bossis et al., 1994; Davis, 1992; Friedberg, 1992). The (110) plane is a closed-packed plane and remains invariant under the BCT → FCC transformation. Together with the constraint of $c=2R$, the transformation path can be described by the equation $b^2 = 12R^2 - a^2$. The hard sphere condition ($a \geq 2R$, $b \geq 2R$) requires $1 \leq a/c \leq \sqrt{2}$, with the limits corresponding to FCC structures oriented differently with respect to the H-field. We performed energy calculations along this path, and the results are shown in Figure 6 for different values of H-field strengths with a fixed applied E-field of 2 *kV/mm*. We find good quantitative agreement with the experiments by using $|\vec{m}_i| = 1.0 \times 10^{-6}$ *emu*. Thus, the only parameter used in the calculation agrees well with the value estimated from the Ni coating thickness.

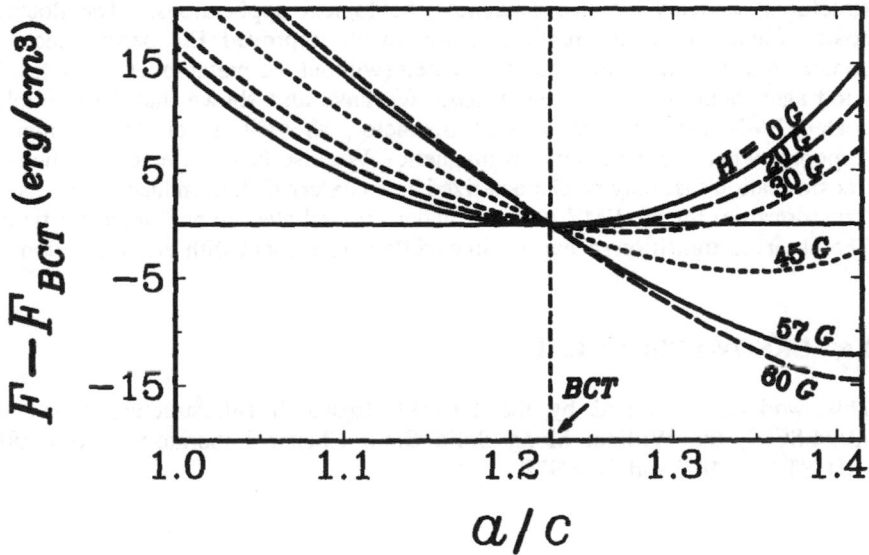

Figure 6. Calculated sum of the electrostatic and magnetostatic free energy densities, minimized with respect to the c and b axes values, are plotted as a function of the a/c ratio. The energy of the BCT structure is used as reference.

From Figure 6, we see that for H-fields below ~ 30 G, BCT remains the optimal structure. This agrees with our observation that there is a reversible regime in the dielectric constant measurement for fields somewhat below 30 G. For H-fields above 30 G, the minimum free energy state rapidly moves to the

FCC structure. This coincides with the observation of non-BCT structures at these H-fields; the observed BCT structure could be due to defects and the lack of temperature effect to overcome local metastabilities. At 57 *G*, FCC is a stable minimum free energy state, as compared to the observation of the clean FCC state at 54 *G*. We note that the columnar structure breaks the cubic symmetry of the FCC structure with $a/c=1$ or $\sqrt{2}$, because the demagnetization effect favors the deformation where a/c increases. The fact that a increases under the action of the H fields implies a decrease of $\overline{\varepsilon}_{yy}$. This is indeed observed as shown in Figure 6. Based on effective medium theory, we estimate that $\overline{\varepsilon}_{yy}$ should decrease by about 0.3% to 5.5% (depending on the shape of the columns and their orientation) as the internal structure changes from BCT to FCC. The observed decrease of about 1% falls within this range.

1.5 CONCLUSIONS

In conclusion, guided by first-principle calculations, we have fabricated structured particles for electromagneto-rheological applications. The doubly-coated (dielectric-metal) particles show much improved ER performance as compared to the un-coated or singly coated (without the metal layer) particles, in good agreements with the calculations. We have also shown that the multiply-coated EMR particles can exhibit magneto-, as well as electro-rheological responses when the metal layer is magnetic. We also have observed a structural transformation that may be characterized as an external-field-induced Martensitic transformation for the EMR particles under crossed electric and magnetic fields. The observed transition scenario is in excellent agreement with the calculations.

1.6 ACKNOWLEDGEMENT

This work is supported by the HKUST Research Infrastructure Grant RI 93/94.SC09, the William Mong Solid State Cluster Laboratory, and CERG HKUST 6142/97P and 6136/97P.

1.7 REFERENCES

Bergman, D. J., 1978, *Phys. Rep.*, **43**, pp. 377.
Bergman, D. J., 1979, *Phys. Rev. B*, **19**, pp. 2359.
Bossis, G., Clercx, H., Grasselli, Y. and E. Lemaice, E., 1994, in *Electrorheological Fluids*, edited by R. Tao and G. D. Roy, World Scientific, pp. 153.
Briker, C. J. and Scherer, G. W., 1990, In *Sol-gel Science: The Physics and Chemistry of Sol-gel Processing*; Academic Press: New York.

Clercx, H. and Bossis, G., 1993, *Phys. Rev. E* **48**, pp. 2721.

Davis, L. C., 1992, *Phys. Rev. A* **46**, pp. R719.

Friedberg, R. and Yu, Y. K., 1992, *Phys. Rev. B* **46**, pp. 6582.

Goldstein, G., 1990, *Mech. Eng.*, **112**, pp. 48.

Havelka, O. K. O and Filisko, F. E., 1995, *Progress in Electrorheology*, (Plenum Press, New York).

Koyama, K. 1996, in *Electro-Rheological Fluids and Magneto-Rheological Suspesions and Associated Technology* (ed. W.A. Bulloguch, World Scientific, Singapore), pp. 245.

Ma, H., Wen, W., Tam, W. Y. and Sheng, P., 1996, *Phys. Rev. Lett.* **77**, pp. 2499.

Milton, G. W., 1980, *J. Appl. Phys.*, **52**, pp. 5286.

Milton, G. W., 1981, *Appl. Phys. A*, **26**, pp. 1207.

Severin, J. W., Hokke, R., Venderwel, H. and G. deWith, G., 1983, *J. Electrochem. Soc.* **140**, pp. 682.

Tam, W. Y., Yi, G. H., Wen, W., Ma, H., Loy, M. M. T. and Sheng, P., 1997, *Phys. Rev. Lett.*, **78**, pp. 2987.

Tao, R. and Sun, J. M., 1991a *Phys. Rev. Lett.* **67** , pp. 398.

Tao, R. and Sun, J. M., 1991b *Phys. Rev. A* **44** , pp. R6181.

Tao, R. 1998, *Phys. Rev. E* **57,** pp. 5761.

Wen, W., Wang, N., Ma, H., Lin, Z., Tam, W. Y., Chan, C. T. and Sheng, P., 1998, preprint.

19 Energy Barrier in the Quantum Nanomagnet: Mn$_{12}$

X. X. Zhang
Department of Physics, The Hong Kong University of Science and Technology, Clear Water Bay, Hong Kong

ABSTRACT

The temperature dependence of the effective energy barrier in the Mn$_{12}$ nanomagnets has been studied with Ac-susceptibility measurements. The results indicate clearly that the effective energy barrier is temperature dependent when the system is in-resonance and independent of temperature when off-resonance, which is a dramatic demonstration of the (thermally assisted) resonant nature of the magnetisation reversal in this system.

INTRODUCTION

Resonant tunneling of magnetisation (or spins) between spin states in magnetic molecules has attracted much attention since it was observed in Mn$_{12}$ acetates (or Mn$_{12}$) several years ago (Friedman *et al* 1996, Hernandez *et al* 1996 and Thomas *et al* 1996). The most interesting feature is that at low temperatures the relaxation rate shows maximums at the regularly spaced magnetic field applied along the easy axis of magnetisation which correspond to resonances between spin states (Hernandez *et al* 1996, Hernandez *et al* 1997 and Luis *et al* 1997). The spin dynamics in Mn$_{12}$ can be described quantitatively by a sample Hamiltonian: $H=-DSz^2-g\mu_BS.B$, D is the magnetic anisotropy energy and positive, indicating the spin of molecule to lie along the Z-axis (the c-axis of the Mn$_{12}$ tetraganal lattice) and B is the magnetic induction. The system can be modelled as a double well potential, as shown in Fig. 1, in which the levels represent the different eigenstates of S$_z$. When magnetic field is zero, all the levels in the double-well match each other (or in resonant), Fig. 1a. The magnetic field (or B) applied along the z-axis will tilt the potential, Fig.1b, raising and lowering the energies of the states, and bringing the levels into resonance at particular value of B ($=nD/g\mu_B$ =nHo, n=0, ±1, ±2, ... (Hernandez *et al* 1997)), Fig.1b. It has been shown that the tunneling rate from the excited level near the top of the barrier is much faster than from the low-lying levels (Hernandez *et al* 1997). The dynamics of Mn$_{12}$ can be described as: the system is first thermally activated to some high-lying levels near the top of the barrier where the tunneling rate is approximately equal to the thermally population of the level(s); it then tunnels through the barrier and decays down into the other well (Fig.1b). Thus, the energy barrier involved in the process should be the effective barrier rather than the total barrier. This effective barrier should vary with temperature when applied magnetic field equals $nD/g\mu_B$. Only when applied field is far from the resonant field H= $nD/g\mu_B$ where the levels in the double-well are fully mismatched, should the barrier be given by classical reduction of energy barrier due to applied field: $U(H)=DS^2(1-H/H_{an})^2$ and is independent of temperature, where $H_{an}=2DS/g\mu_B$ is the anisotropy energy. This has not been experimentally demonstrated yet and the energy barrier U ($=DS^2$, S=10) between 56K to 86K was extracted from different experiments using different techniques (Luis *et al* 1997, Sessoli 1995 and Sessoli *et al* 1993). Here we report on the experimental evidence of the temperature dependent effective energy barrier when the applied field equals the resonant field (H= $nD/g\mu_B$) or near to it and the temperature independent classical energy barrier when applied field is far from the resonant field H= $nD/g\mu_B$.

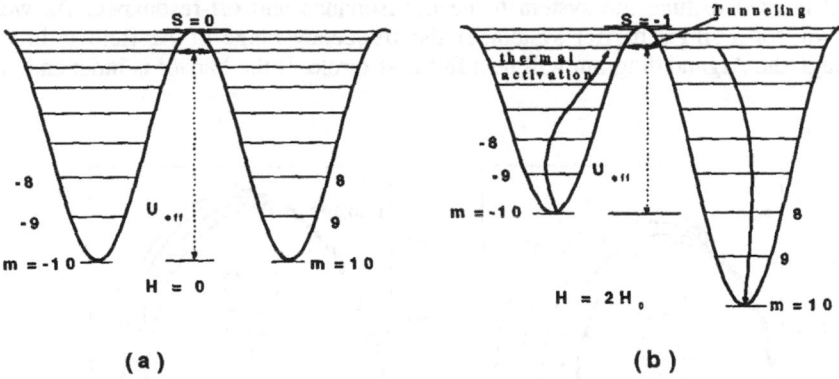

(a)   (b)

Fig. 1 Schematic diagram of thermally assisted resonant tunneling process.

RESULTS AND DISCUSSION

The sample used in this work is the magnetic field aligned small crystallites embedded in the epoxy. Magnetic characterisation has been carried out on a commercial Quantum Design SQUID magnetometer. Magnetic hysteresis loops measured below 3K show regularly spaced jumps with a step of $H_0=0.44$T, in agreement with the previously reported data (Friedman *et al* 1996, Hernandez *et al* 1996, Thomas *et al* 1996 and Hernandez *et al* 1996). The magnetic data suggest a misorientation no more than 1°.

Relaxation rate of magnetic moment in a single domain particle with uniaxial anisotropy is given by $\Gamma=\Gamma_0\exp(-U/k_BT)$, where U is the anisotropy energy barrier and Γ_0 is the attempt frequency. The Ac-susceptibility of a system composed of identical particles is given by:

$$\chi'=\chi_0/(1+\omega^2\tau^2) \quad \text{and} \quad \chi''=\chi_0\omega\tau/(1+\omega^2\tau^2) \tag{1}$$

Where χ_o (=$(g\mu_B S)^2/T$, for Mn_{12}) is the dc superparamagnetic susceptibility, ω is the angular frequency of applied Ac field; $\tau = \Gamma^{-1} = \Gamma^{-1}_0 \exp(U/k_B T)$, the relaxation time of the magnetic moment. It is well known that when $\omega\tau = 1$, that is, $\omega = \tau^{-1} = \Gamma = \Gamma_0 \exp(-U/k_B T)$, χ'' shows a maximum at the blocking temperature T_B. By measuring temperature dependent susceptibility with different Ac field frequencies in zero dc-field, different blocking temperatures can be measured. Energy barrier U and attempt frequency Γ_0 are, therefore, easily extracted in fitting the obtained blocking temperatures to $\omega = \tau^{-1} = \Gamma = \Gamma_0 \exp(-U/k_B T)$. In above process, the energy barrier has been assumed to be temperature independent. This method has been used to extract the energy barrier in the Mn_{12} (Luis *et al* 1997, and Sessoli 1995), which gives a effective energy barrier being smaller than the whole barrier $U = DS^2$ as discussed above. It is found that the data can be fitted well to $\omega = \tau^{-1} = \Gamma = \Gamma_0 \exp(-U/k_B T_B)$, and no irregularity has been found. We will demonstrate below that the energy barrier is temperature dependent when the system is in resonance. Fig. 2 shows the Argand diagrams obtained at different temperatures in different magnetic fields applied along easy axis. The applied field is used to tune the system to be in-resonance and off-resonance. As well known, for a single barrier system, if the frequency range of Ac-field is broad enough the Argand diagram will be a full half-circle. If the barrier is independent

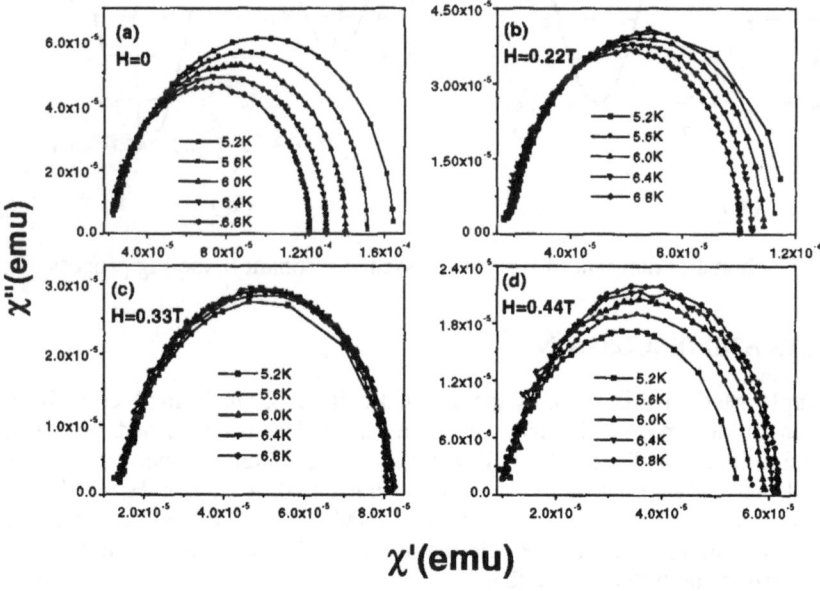

Fig.2 Argand diagrams obtained with different applied dc-field.

of temperature, the Argand diagram measured at different temperatures will collapse into one master curve. It is clear seen that when $H_{dc}=0$, that is, the system is in-resonance, the curves scatter and no master curve is found, which must be caused by the temperature dependent energy barrier. As discussed above, in Mn$_{12}$ the energy barrier is determined by the level(s) whose tunneling rate is comparable to its thermal population rate when the system in resonance. The thermal population rate of a particular level depends on the energy difference between this level and bottom of the well and the temperature $\Gamma_p \propto \exp(-U/k_B T)$. At low temperatures the population rate of a level near top of well is very small, the low-lying level(s) may dominate. Since the thermal population increases exponentially with temperature, the high-lying level(s) begins to dominate with increasing temperature. The shift of the dominating level results in a change of effective energy barrier when system is in-resonance. When the applied field is far from resonant field, that is the system is off-resonance, the barrier must not be temperature dependent. Therefore, the Argand diagrams for different temperatures must collapse into on curve, which is clearly seen in Fig. 2c for field of 0.33T. With increasing field from 0.33T, the curves begin to deviate (or scatter), because the field is approaching the resonant field (Fig. 2d). The total energy barrier can be obtained from the susceptibility data measured with applied field of 0.33T using $U=U_0(1-H/H_{an})^2$ and $H_{an}=8.8T$ (Ho=H_{an}/2S). The extract barrier, $U_0=DS^2=87.2K$ is much larger than the effective energy barrier when the system in resonance and in agreement with $U=DS^2=86K$ obtained from the EPR measurements (Luis *et al* 1997 and Sessoli et al 1993).

ACKNOWLEDGEMENT:

This work is supported by RGC (Hong Kong) HKUST6111/98P

REFERENCES

Friedman J.R., Sarachik M.P., Tejada J. and Ziolo R.F., 1996, Physical Review Letters, 76, pp.3830-3.

Hernandez J. M., Zhang X.X., Luis F., Bartolome J., Tejada J., Ziolo R., 1996, Europhysics Letters, 35, pp.301-306.

Hernandez J. M., Zhang X. X., Luis F., Tejada J., Friedman J.R., Sarachik M.P., Ziolo R., 1997, Physical Review B55,pp5858-5865

Luis F., Bartolome J., Fernandez J. F., Tejada J., Hernandez J.M., Zhang X.X., Ziolo R., 1997, Physical Review B55, pp.11448-11456

Sessoli R., 1995, Molecular Crystals & Liquid Crystals, 273-274, pp.801-813.

Sessoli R. Tsai H. L., Schake A. R., Wang S. Y., Vincent J. B., Folting K., Gatteschi D., Christou G. and Hendrickson D. N., 1993, Journal of the American Chemical Society,115, pp 1804-1816

Thomas L, Lionti F, Ballou R, Gatteschi D, Sessoli R, Barbara B., 1996, Nature, 383, pp.145-147.

20 The Percolation Model: the Tribological Mechanism of Nano Particle Filled Polymer Composites

Feng-yuan Yan and Qun-ji Xue
Laboratory of Solid Lubrication, Lanzhou Institute of Chemical Physics, Chinese Academy of Sciences, Lanzhou 730000, P. R. China

ABSTRACT

A method for preparation of nano ZrO_2 in liquid medium and for simultaneous mixing of the resultant nano ZrO_2 with powder of polytetrafluoroethylene (PTFE) was established. The tribological behavior of nano ZrO_2-PTFE composites was investigated under dry friction. It was found that at lower filler content, the optimal filler content for minimum wear rate of nano ZrO_2-PTFE changed with the particle size. The larger the filler size, the higher the optimal filler content. However, at higher filler content, the critical filler content was almost a constant and had no relation with the particle size. Combined the tribological behavior with the microstructural characteristics of the composites, a percolation model was proposed to explain the experimental results. It was concluded primarily that the wear behavior of the composites could be mathematically treated as a special percolation phenomenon. The critical content of the nano particles with different size in matrix for the lowest wear rate could be correlated with the composite system's percolation threshold, where the critical matrix ligament thickness was about 11.64 nm, and the critical volume fraction at percolation threshold 0.21.

1 INTRODUCTION

In recent years there has been growing interest in the use of polymers and polymer based composites in tribological situations. This kind of interest has stimulated scientific research on new formulations of polymer based composites for tribological applications (Bahadur and Kapoor, 1992). Comparing with the tribological actions of micrometer (μ)-sized fillers, researchers had found that some nano particles used as fillers were very effective in modifying the tribological behaviors of polymers at lower filler content (Wang, *et al.*, 1996). This was attributed to the chemical activities of nano fillers. But why and at what conditions the nano fillers were more effective than the μ-sized fillers in modifying the tribological properties of polymers, and if there was some relation between the optimal filler proportion with particle sizes are unclear.

The brittle-ductile transition mechanisms of polymer based composites have been successfully interpreted in terms of percolation concept (Wu, 1992, Fu and Wang, 1993). Since the tribological process was closely related to the brittle, tough and other mechanical properties, it was expected that deep understanding of the tribological mechanisms of polymer composites could be realized by use of the percolation theory.

2 EXPERIMENTAL

The coagulation of nano particles in polymer matrix was usually inevitable because of their high chemical activities. In this paper, a sol-gel method for preparation of nano ZrO_2 in liquid medium and for simultaneous mixing of the resultant nano ZrO_2 with PTFE powder was established. The mixture of PTFE powder of diameter about 50 μm with certain content of dispersant and the analytically pure $ZrOCl_2 \cdot 8H_2O$ was simultaneously dispersed in distilled water under continuous stirring. Then the mixture was antagonized to $ZrO(OH)_2$ gel by adding excess ammonia gradually. The chemical reaction is as below:

$$ZrOCl_2 + NH_4OH \rightarrow ZrO(OH)_2\downarrow + NH_4Cl$$

Since the pyrolytic products of $ZrO(OH)_2$ above 380 ℃ are ZrO_2 and H_2O, so PTFE-ZrO_2 composite could be obtained by heat compression molding of the dried mixture of $ZrO(OH)_2$/PTFE. If ZrO_2 particle was ball-shaped, the average size of nano ZrO_2 could be evaluated from the specific surface area of the powder mixture. The size of ZrO_2 could be controlled by changing the heat compression molding time.

The friction and wear tests were conducted on a ring-on-disk friction and wear tester at the speed of 0.419 m/s, load 200 N, test duration 2 hours (distance 3016m), room temperature 20 ℃, and the couple material of stainless steel.

3 RESULTS AND DISSCUSSION

3.1 The Tribological Behavior of Nano ZrO_2-PTFE Composites

Figure 1 shows the wear rate variations of composites along with the filler content and the particle sizes. It is seen that for the composites with particle size below 97.6 nm, two critical contents corresponding to the lowest wear rate are observed, while the composites with particle size of 97.6 nm have only one. At lower filler content, the smaller the particle size, the lower the critical content and the higher the wear rate. At the higher particle content, the critical content and the wear rate are almost constants which have no relation to the particle size.

By comparing the particle sizes of ZrO_2 with PTFE, it was estimated that the coagulation of nano ZrO_2 occurs inevitably above a certain filler content, the smaller the particle size, the more serious the coagulation. The critical content for coagulation of nano particle is usually smaller then 10 wt.% or 4 vol.%. Based on the tribological test results, it was supposed that the critical content for optimal wear behavior at higher content is correlated to the coagulated nano particles in the composites. The size of the coagulated ZrO_2 was estimated to be about 100 nm.

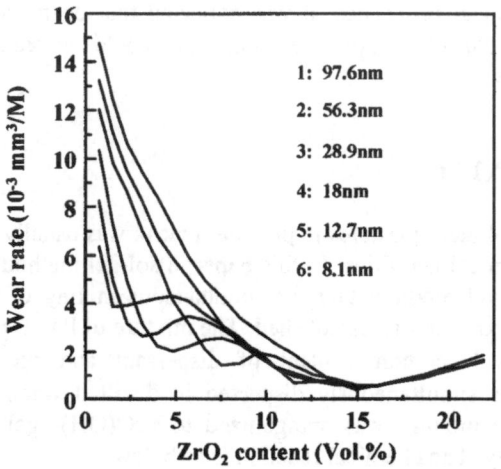

Figure 1 The wear rates of PTFE based composites filled with nano ZrO_2 of different content and size

3.2 The Tribological Percolation Model

Particles dispersed in a matrix forming divided stress volumes were supposed to distribute randomly, that is, the equal-sized particles occupying a random lattice. As the divided stress volumes are interconnected, allowing frictional damage process on surface or sub-surface to pervade over the entire matrix in the deformation zone, the best wear resistance corresponding to a tribological percolation threshold could be obtained. Figure 2 shows the figured tribological percolation model (Marked B).

A stress volume centers at a filler particle and includes a concentric annular shell of the matrix of constant thickness $\tau_c/2$. The diameter of the stress volume S is as $S = d + \tau_c$, where d is the average particle diameter. Two neighboring stress volumes are interconnected when L≤S, where L is the center-to-center interparticle distance. It relates to the mutual contact of the stress volume at percolation threshold (L=S) and to the point of the lowest wear rate, where $S_c=d_c+\tau_c$; subscript c denotes the critical condition. The stress volume fraction Φ_s is: $\Phi_s = \Phi_r (S/d)^3 =$

$\Phi_r [(d+\tau_c) / d]^3$, where Φ_r is the particle volume fraction. At percolation threshold condition, the critical stress volume fraction Φ_{sc} can be obtained: $\Phi_{sc}=\Phi_{rc}(S_c/d_c)^3$.

According to the basic theory of percolation, if the wear behavior of the composites is a percolation phenomenon, two scaling laws should be satisfied. First, the percolation threshold for stress volumes should be constant, independent of particle volume fraction. For nano ZrO_2-PTFE composites, Φ_{sc} was found to be a constant about 0.21, where τ_c was about 11.64 nm. Secondly, the wear rate should obey the near-threshold critical scaling law: $W=A(\Phi_s-\Phi_{sc})^g$, where W is the wear rate, A the prefactor, $(\Phi_s-\Phi_{sc})$ the excess critical stress volume and g the critical exponent. The g for ZrO_2-PTFE is about 1 by plotting log W versus log $(\Phi_s-\Phi_{sc})$.

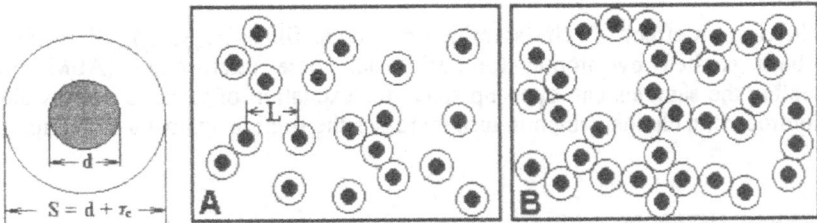

Figure 2   Schematics of a stress volume and illustration for the
continuum percolation of stress volumes

4   CONCLUSIONS

As nano particles possess higher physical and chemical activity and strong coagulating tendency, nano ZrO_2-PTFE composites usually have two critical content for the lowest wear rate. At the lower filler content the critical content corresponds to the real nano particles dispersed composites, while at higher filler content to the coagulated particles filled composites.

The tribological behavior of nano ZrO_2 filled PTFE composites can be considered as a percolation phenomenon. The correlation between the wear rate, particle size and particle content well satisfied with the percolation scaling laws.

5   REFERENCES

Bahadur, S. and Kapoor. A., 1992, The effect of ZnF_2, ZnS and PbS fillers on the tribological behavior of nylon 11, *Wear*, **155**, pp. 49~61.

Wang, Q.H, *et al.*, 1996, The effect of particle size of nanometer ZrO_2 on the tribological behaviour of PEEK, *Wear*, **198**, 216-219.

Wu, S.H, 1992, Control of Intrinsic Brittleness and Toughness of Polymers and Blends by Chemical Structure: A Review, *Polymer International*, **29**, 229-247.

Fu, Q. and Wang, G.H., 1993, Polyethylene Toughened by $CaCO_3$ Particles-Percolation Model of Brittle-Ductile Transition in HDPE/$CaCO_3$ Blends, *Polymer International*, **30**, 309-312.

21 Tribological Properties of Self-assembled Monolayers

Xudong Xiao, Chun Zhang and Bing Wang
The Hong Kong University of Science and Technology, Hong Kong

ABSTRACT

The lubrication effect of self-assembled molecular films, $CH_3(CH_2)_{n-1}SH/Au(111)$, has been studied by atomic force/frictional force microscopy (AFM/FFM). Annealing the samples can open up space for excitation of Gauche defects along the hydrocarbon chains, which in turn increases the friction against a Si_3N_4 tip.

1. INTRODUCTION

The frictional property at atomic scale has become an interesting topic in recent years mainly due to the invention of atomic force microscope (AFM) and its offspring frictional force microscope (FFM). Together with the surface force apparatus (SFA) and the quartz-crystal microbalance (QCM) methods, fundamental origins of the tribological phenomena, including adhesion, friction and wear, at the atomic scale can be investigated [1]. On the application side, the high density data storage systems and the recently developed micromachine systems put more and more stringent requirement on the thin film lubrication, and push it to the monolayer thickness limit. It is thus curious to investigate how the structure of a molecularly thin film such as self-assembled monolayers (SAMs) influences the lubrication. At this scale, the viscosity which plays an important role in the traditional liquid lubricants may become irrelevant. Rather, the ordering of the monolayer may determine the lubrication effect.

In this paper, we report an experimental study of alkylthiols, $CH_3(CH_2)_{n-1}SH$ with n=7, 10, and 18, self-assembled on Au(111). From scanning tunnelling microscopy (STM) studies, it is known that these molecules form ordered $(\sqrt{3} \times \sqrt{3})R30°$ structure after self-assembly at room temperature. With annealing at sufficiently high temperatures for an extended time period, the ordering of the films would first degrade and then form another ordered phase – stripe phase as a result of partial desorption. Thus, these monolayers with different amount of annealing provide us a playground to test the ideas on how tribological properties are related to the film structures. Our friction results on these films measured by FFM with a Si_3N_4 tip show that friction increases as the monolayer ordering deteriorates and then reaches to a saturation level with ~4 fold increase for the stripe phase. The increase of the friction as the monolayer ordering degrades can be understood as due to the increasing accessibility to Gauche kink excitation in the hydrocarbon chains by the AFM tip, which opens up an energy dissipation channel. This argument seems to hold equally well for the ordered stripe phase, for which the molecular density is lower than those of the disordered intermediate phases. Thus, we conclude that lubrication effect of these monolayers does not simply depend on whether the films are ordered, rather it depends on whether the films allow access to additional energy dissipation channels.

2. EXPERIMENT

The $CH_3(CH_2)_{n-1}SH$ (n=7, 10, and 18) self-assembled monolayer films on Au(111) were made by immersing the freshly evaporated gold substrates into a 2 mM solution for 48 hours [2]. The annealing of the samples was carried out in a glove box purged with dry air to a relative humidity level below 10%. For each annealing temperature, a fresh sample was used. The annealing effect of the films at a given temperature for a given period of time was often monitored by STM. The AFM/FFM used in the experiment was home-made and employed a laser deflection detection system. It was operated at ambient conditions. Both the friction and the topography imaging measurements were taken by a commercial Si_3N_4 cantilever/tip with a nominal force constant of 0.5 N/m.

3. RESULTS AND DISCUSSION

In Figure 1, the representative friction-versus-load curves for self-assembled alkylthiol films, $CH_3(CH_2)_{n-1}SH$, with different chain length before and after annealing at various temperatures for 10 hours are shown. For n=7, annealing the monolayer at 60°C results in negligible change in friction. However, when the annealing temperature is 70°C, a significant increase of friction by about a factor of 3 is observed. Annealing at higher temperature (80°C) seems to cause no further change. For n=10, an annealing temperature of 70°C already induces a change in friction, but the effect seems to saturate only if annealed above 80°C. The increase of friction is now about four folds. For n=18, the annealing temperature must be ~100°C in order to induce a comparable change in friction. Note that the friction increase after annealing not only occurs in the slope but also in the magnitude of the offset. Comparing to the fresh films, the friction at the snap-into-contact (zero load) point for the annealed films is significantly larger. As the chain length increases, this friction value seems to increase as well. For all the films, AFM topographic images have been taken. Consistently, well-ordered, lattice-resolved images with $(\sqrt{3} \times \sqrt{3})R30°$ structure are observed for fresh films. As long as the friction signal remains comparable to that of the fresh samples, annealing does not degrade the topographic images. For example, after annealing the $CH_3(CH_2)_{17}SH$ sample at 90°C for 10 hours, topographic images still show the $(\sqrt{3} \times \sqrt{3})R30°$ structure with little degradation. Once the annealed samples show the saturation friction, no more lattice-resolved AFM images can be obtained. However, for these samples, ordered stripe phase can be clearly identified in the STM images. From the above observations, we conclude that i) annealing alkylthiol films at sufficiently high temperature an induce an irreversible change to the films; ii) the annealing temperature required to cause the friction change increases as the chain length; iii) the increase in friction is well correlated with the

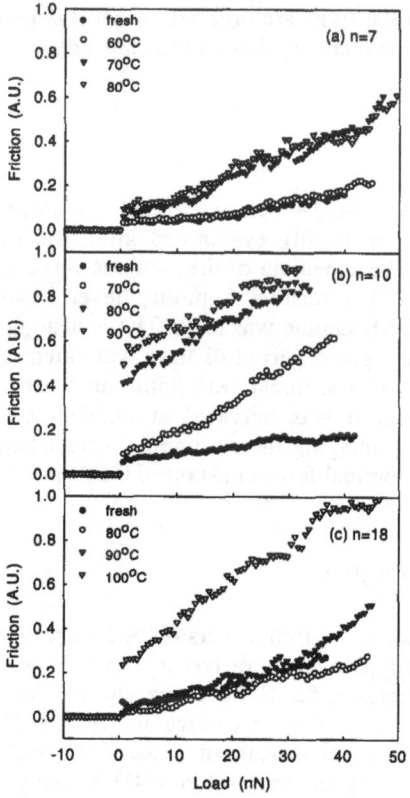

Figure 1. Representative friction-versus-load curves for alkylthiol, $CH_3(CH_2)_{n-1}SH$, monolayers self-assembled on Au(111) before and after annealing at different temperatures. The annealing time for each sample is 10 hours and the annealing temperatures are indicated in the figure. (a) n=7, (b) n=10, and (c) n=18.

disappearance of the lattice-resolved AFM images, or with the appearance of the stripe phase as seen by STM.

In Figure 2, we further show the friction results of the $CH_3(CH_2)_9SH$ film as a function of annealing time at 75°C. As the annealing time increases, the film starts to disorder to certain extent as observed by STM. Accompanied with the disordering, the friction of the film increases. With 1-2 hours annealing, the disordering is very minimal and the friction increases only at high loads (>30nN). When the annealing time extends to 3-6 hours, the disordering increases further and now the friction increases even at low loads. With ~ 10 hours annealing, the stripe phase appears to have completed and the friction reaches a saturation level.

From the above results, we have seen that friction of the akylthiol films does not simply depend on whether the film is ordered. As the density of the film is

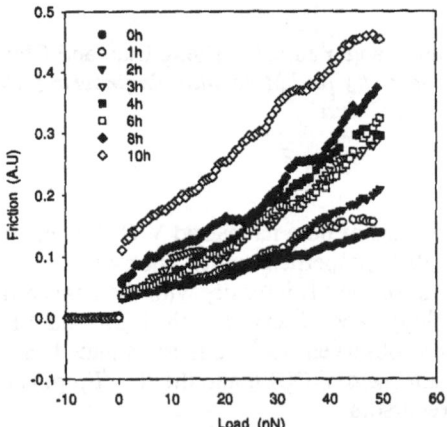

Figure 2. Representative friction-versus-load curves for alkylthiol, $CH_3(CH_2)_9SH$, monolayers for different annealing period at temperature 75°C.

lowered, the friction increases. This can be explained as follows. With the reduced compactness of films by partial desorption, the van der Waals interaction among the hydrocarbon chains decreases, and therefore the chain may no longer retain a trans-conformation, in particular under applied load. In another word, Gauche kinks can be excited in the remaining molecules as the tip scans over the film, which opens up an additional energy dissipation channel. This is why annealing of the films at sufficiently high temperatures causes reduction of lubrication.

4. CONCLUSION

In summary, we have investigated the annealing effect on lubrication for a number of alkylthiol films. The results have shown that friction increases as the probability for Gauche defect excitation increases independent of the ordering of the films.

REFERENCES

[1] Bhushan, B.; Israelachvili, J. N.; Landman, U. *Nature* **1995**, 374, 607.
[2] Sondag-Huethorst, J. A. M. ; Schonenberger, C.; and Fokkink, L. G. J. *J. Phys. Chem.* **1994**, 98, 6826.

22 Nanosized Rare Earth Activated Phosphors

Lingdong Sun, Jiang Yao, Chunsheng Liao and Chunhua Yan
College of Chemistry and Molecular Engineering, Peking University, Beijing 100871, China

ABSTRACT

Recent success in synthesis of pure nanosized Y_2O_3:Eu enables us to evaluate their photoluminescence (PL) properties under ultra-violet photoexcitation and to explore the relationship of the PL intensity with the concentration of the activator. Complicated morphology was observed with TEM and SEM and no real free-standing particles were observed. Eu^{3+} ions are situated in a strong electric field environment in the nanosized Y_2O_3:Eu and the $^5D_0$-$^7F_2$ transition intensity depends on [G]/[RE] in the precursors.

1. INTRODUCTION

Nanosized materials have been studied vigorously in recent years because of its significant properties differing from the bulk. Most of the research works are devoted to optical properties induced by the quantum confinement. Although the nonradiative recombination routes brought by the surface states are dominant in most of the confined systems, efforts were made to decrease the dangling bonds by surfactant. Bhargava reported another way to achieve the recombination not dominated by the defects. He incorporated an impurity to a confined system, and found the recombination route can be transferred from the defects to the impurity states and the efficiency is increased. By changing the impurity ions, the emission band can be adjusted. This leads to the study of nanosized luminescent materials.

Rare earth (RE) ions, especially Eu^{3+}, Tb^{3+} ions are good activators for luminescence materials and Y_2O_3:Eu is widely used as CTR and tricolor red phosphors. Furthermore, nanosized Y_2O_3:Eu have significant promise in FEDs. In this paper, we report a simple large-scale synthesis of nanosized Y_2O_3:Eu by glycine activated combustion method and the results that the particles linked up to exhibit three-dimension porous structure were proposed. Detailed characterization of the size, morphology, composition, and structure were carried out. The characteristic optical properties were studied as well.

2. EXPERIMENTAL

The appropriate rations of RE (Y^{3+}, Eu^{3+}) nitrate and glycine solutions were mixed together as the precursors. Light yellow precursor solutions were evaporated by heating until dense and melted salt with brown was got. Bubbles rose from the mixture and at last the combustion occurred. The reaction last for seconds and the volume expand seriously while the foamy samples produced. The brown foggy, NO_2 gas was also observed. The resultant is crisp and brittle.

X-ray diffraction (XRD) was carried out on Rigaku RINT Dmax-2000 with Cu-Kα irradiation with X-ray generates at 40 kV/100mA. The morphology and composition analysis were performed by means of Hitachi H-9000 NAR high-resolution transmission electron microscope (HRTEM) operating at 300 kV equipped with an EDAX detector. Amary field emission scanning electron microscope (SEM) was also used for morphology analysis. Photoluminescence (PL) measurement was performed at room temperature using a Hitachi F-4500 spectrophotometer.

3. RESULTS AND CONCLUSIONS

3.1 Structure and Morphology of the nanosized Y_2O_3:Eu

It is known that much glycine can provide much more energy and increase the system temperature. Fig. 1 is the XRD patterns for three samples Y1, Y2 and Y3, they correspond to the concentration ratio of glycine and RE, [G]/[RE], of 1, 1.3 and 1.7, respectively. Less glycine makes the diffraction line broadened and this comes from the disorder of the system and the size distribution. Increasing glycine in the precursors leads to perfect crystallization of the particles. For the as-prepared products, further low temperature (less than 500°C) anneal will also improve the crystallization.

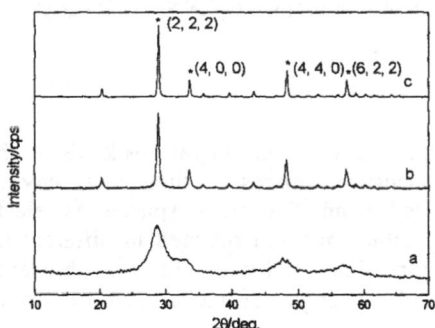

Figure 1. XRD patterns for sample Y1, Y2 and Y3 (correspond to a, b and c respectively).

From the state of the resultant, we guess that the ultra fine particles are produced. For characterizing the nanosized materials, the best and direct way is to obtain the size and morphology under the TEM. Fig. 2(a) is a typical TEM micrograph of the samples. We can't identify the isolated particles from the network as expected. EDAX was used to get the composition of the network. It is found that Y, Eu and O elements are co-existed as prepared. The edge of the network is not as smooth as the nanowires and it is circled to form hollow structures with different size of pores. Different [G]/[RE] did not have evident change in the morphology. The network diameter increased from 5nm to about 7nm for [G]/[RE] of 1 and 1.7. Fig. 2(b) is the ED pattern. The diffraction rings indicate the small particles. As the magnification increased, lattice fringe with different lattice spacing is visible (Fig. 2(c)), and some of them are crossed by sets of lattice fringe. The lattice spacing corresponds to different planes indicated that the prepared nanocrystals have different growth directions. Fig. 2(d) is a typical SEM image of the as-prepared products, the inside bar is 1 µm. The fragment of the network was shaped by the impingement of the explosive gas to form the three dimensional porous structure. As the glycine increased in the reaction, the fragment tends to be more porous as shown by the SEM image.

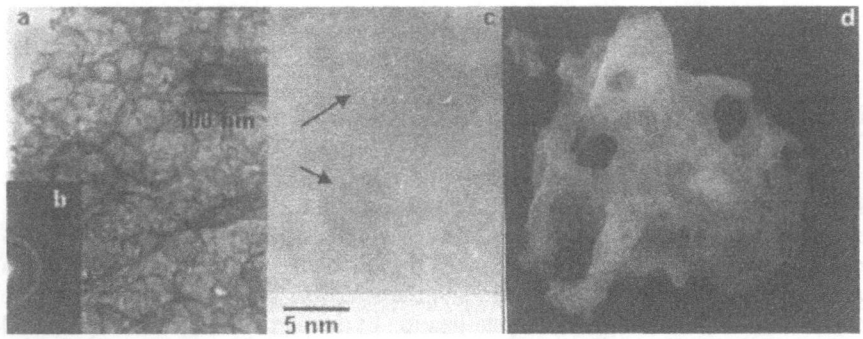

Figure 2 (a) and (b) are TEM micrograph and ED pattern of sample Y1, (c) HRTEM indicates the particle size is about 5 nm and (d) is typical SEM micrograph.

3.2 PL properties

Rare earth ions used as the activators in various kinds of phosphors and Eu^{3+} ions are also served as sensitive probes of their local environment. The outer $4f$ electrons are shielded from the outer spaced $5s$ electrons and the lattice perturbations. The f orbit transition retained in different kinds of host, but it is sensitive to local crystal fields and will split into sub-energy levels. The splitting pattern depends on the strength of the crystal field and the local symmetry environment.

The as-prepared sample is excited with a xenon lamp in the ultraviolet region and the corresponding room-temperature PL spectra are shown in Fig. 3. Lowering the [G]/[RE] value to 1, broadened spectra were observed from the disordered system. Increasing the [G]/[RE] value, three distinct $^5D_0$-$^7F_1$ transition can be clearly detected, that indicates that the local symmetry of the Eu^{3+} ions must be of a lower group, and the splitting of the peaks (150–170 cm^{-1}) indicates relatively large local field.

The ratio between the transition strength of the $^5D_0$-$^7F_1$ and $^5D_0$-$^7F_2$ can be used as the indicator of the local environment. It is stated (Peacock, R.D., 1975) that the f-f transition is electric dipole forbidden. But in some cases, for the $^5D_0$-$^7F_2$ transition, the mixing of asymmetry with other transition makes the forbidden partially permitted and leads to a finite oscillator strength. But for the $^5D_0$-$^7F_1$ transition, the electric dipole is forbidden in any cases, only magnetic dipole transition contributes to the transition intensity. For the as-prepared sample Y1, further anneal will change the local environment, resulting in the $^5D_0$-$^7F_1$ transition intensity increases relative to that of $^5D_0$-$^7F_2$ and it splits into three evident isolated peaks. For sample Y2 and Y3, further anneal increase the $^5D_0$-$^7F_2$ transition but will not change the relative intensity of the sensitive transition.

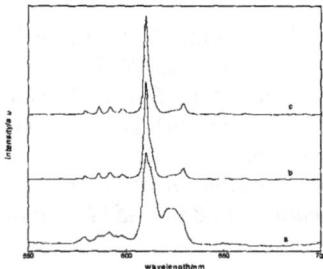

Figure 3 PL spectra of sample Y1, Y2 and Y3 (correspond to a, b and c, respectively).

Previous work report that the quenching concentration is increased for the nanosized Y_2O_3:Eu compared to that of the bulk. The main reason is that the nanostructure introduces much more interface and the energy transfer is blocked, *i.e.*, less energy is depleted through the nonradiative center. From this postulation, the quenching concentration should increase with decreasing particle size. The recent results are contradicted and the real mechanism is under debate. For the combustion method, the quenching concentration increased from ~5 % to 10% for [G]/[RE] from 1 to 1.3. Increasing [G]/[RE] will not increase the quenching concentration any more. Although the relationship is not certain, the quenching concentration of the nanosized material is higher than the corresponding bulk is doubtless. This phenomenon suggests that the large volume-surface ratio induced surface states is not the main factor for nonradiative recombination.

ACKNOWLEDGMENT

This work is supported by Fundamental Research Project and Natural Science Foundation. The authors would like to thank Croucher Foundation, and useful discussions with Prof. Huang Shihua from Changchun Institute of Physics, CAS.

Bhargava, R.N. *et al.*, 1994, Optical Properties of Manganese-Doped Nanocrystals of ZnS, *Physics Review Letters*, 72(3), pp. 416-419.

Tao Ye *et al.*, 1997, Combustion Synthesis and Photoluminescence of Nanocrystalline Y_2O_3:Eu Phosphors, *Materials Research Bulletin*, 32(5), pp.501-506.

Peacock, R.D., 1975, In *Structure and Bonding* (New York: Springer -Verlag), 22(2), pp. 83-122.

23 Properties of Co/Al$_2$O$_3$/Co Tunnel Junction with Co Nanoparticles Embedded in Al$_2$O$_3$

S. F. Lee[1], P. Holody[2], F. Fettar[2], A. Vaurès[2], J. L. Maurice[2], J. Barnas[3], F. Petroff[2], A. Fert[2], Y. Liou[1] and Y. D. Yao[1]

[1]*Institute of Physics, Academia Sinica, Nankang, Taipei 115, Taiwan,*
[2]*Unité Mixte de Recherche CNRS/Thomson-LCR, Domaine de Corbeville, 91404 Orsay, France and*
[3]*Magnetism Theory Division, Institute of Physics, Adam Mickiewicz University, ul. Umultowska 85, 61-514 Poznan, Poland*

ABSTRACT

Magnetic tunnel junctions consist of two Co electrodes separated by Al$_2$O$_3$, in which one or several layers of Co nanoparticles are embedded, are fabricated and studied. The shapes of the Co nanoparticles are shown by TEM to be close to spherical and the blocking temperature is about 30 K. A good magnetic tunnel junction shows a non-linear current-voltage (I-V) curve and the tunneling probability depends on the relative orientation of the moments in the two electrodes. Since the junction has non-linear I-V curves, V-biased and I-biased magnetoresistance (MR) are different. The ratio between I-biased and V-biased MR depends on the curvature of the I-V curve, which in turn depends on the physical properties of the insulating barrier. The existence of Co nanoparticles inside the insulating barrier gives rise to Coulomb blockade effect at low temperature. Our results show MR does not change significantly crossing blocking temperature, but I-biased MR is strongly suppressed at low bias, low temperature.

INTRODUCTION

Electron tunneling through a thin insulating barrier between two ferromagnetic metal electrodes was first reported over twenty years ago (Julliere, 1975). A simple model was proposed based on the difference of conduction electron spin-polarization between the two electrodes, and was later supported by theoretical work (Slonczewski, 1989). Taking P$_1$ and P$_2$ as the conduction electron spin polarization of the top and bottom electrodes, the maximum resistance change occurs when the magnetic moments change from antiparallel to parallel and can be written as:

$$MR = \frac{R_{AP} - R_P}{R_P} = \frac{2P_1 P_2}{1 - P_1 P_2}. \qquad (1)$$

Recently, this field has attracted lots of attention since a report on reproducible and large MR at room temperature (Moodera *et al.* 1995). Several groups have reported similar results (Miyazaki and Tezuka, 1995; Schelp *et al.*, 1997). Despite the advent of thin film fabrication technology, the quality of the insulating barriers on ferromagnetic metal remains the center issue. One common feature of all the tunneling MR is its strong dependence of bias voltage. For a 'good' junction, MR decreases by a factor of 10 from 0 to 0.5V, for a 'better' junction by 60% (Moodera *et al.* 1998). There is no satisfactory explanation by any theories so far (Zhang *et al.*, 1997).

Resistance of bad junctions is small and does not scale with the inverse of junction area. Even a reverse of voltage drop from current direction can occur and the MR can approach infinite (van de Veerdonk *et al.*, 1997). A good junction shows non-linear I-V curve, different I-V curves can be traced out in different external magnetic fields. Derivation can be easily shown that measuring MR with constant voltage and constant current is different and follows the relationship:

$$\frac{MR_{constV}}{MR_{constI}} = \frac{dV / dI}{V / I}. \qquad (2)$$

The above equation indicates that when the I-V curve is linear, there is no difference between how the MR is measured. But when I-V curve is non-linear, MR values can depend strongly on how it is measured.

EXPERIMENTS

We fabricated layered Co nanoparticles inside Al₂O₃ by sputtering from individual targets onto Si substrates sequentially. The shapes of the Co nanoparticles are shown by TEM to be close to spherical, as shown in Fig.1. The sizes are narrowly distributed with Gaussian distribution. Average diameter is 2.7 nm and inter-particle distance is 2 nm. Magnetic properties are measured with Superconducting Quantum Interference Device (SQUID) susceptometer. Field-cooled and zero-field-cooled magnetization curves of these samples show blocking tempera-

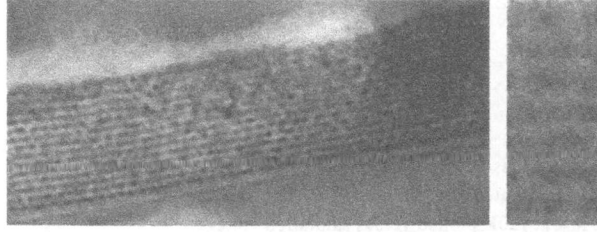
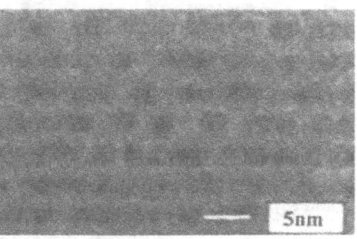

Figure 1 TEM pictures of layered Co clusters in Al₂O₃.

tures around 30K. Co clusters are superparamagnetic above the blocking temperature ferromagnetic below. Magnetic tunnel junctions are produced by sputtering
through contact masks. Both top and bottom electrodes are 15nm Co. The afore-
mentioned Co-cluster embedded Al_2O_3 is sandwiched in between.

To measure DC I-V curves and MRs, we use a Keithley 236 source-measure
unit. In V-biased case, two-point method is utilized because both the voltage source and Ampere meter have small impedance and need to be in series with the sample. This means the voltage drop is across the junction plus two electrode arms. It is
possible to use the other pair of electrodes to measure and control the real voltage
drop across the junction. Unfortunately the current noise is usually much bigger. In
I-biased case, four-point method is utilized. Electrodes' resistance is mostly elimi-
nated but tunneling current is close to uniform only when junctions' resistance is
much larger than electrodes'. We define the MR as in Equation (1), with saturation
resistance as reference.

MR measurements at different voltage or current bias at several temperatures
are summarized in Figure 2. (a) and (b) are data on a sample with single layer of
Co nanoparticles. (c) and (d) are for a sample with 7 layers of Co in the junction.
Notice that there is no obvious change above and below blocking temperature, 30K.
From (a) and (c) we see similar MR decrease with increasing voltage as reported
by other groups. The decrease rate is close to 'better' planar junction (Moodera *et
al.*, 1998). The decrease rate is slightly slowly at higher temperature. (b) and (d)
show more complex I-bias behavior. We plot in (a) I-biased MR versus saturation
voltage at 120K. At low bias, when I-V curve is close to linear, two different ways
of measuring MR almost equal. I-biased MR decreases faster at high bias. This

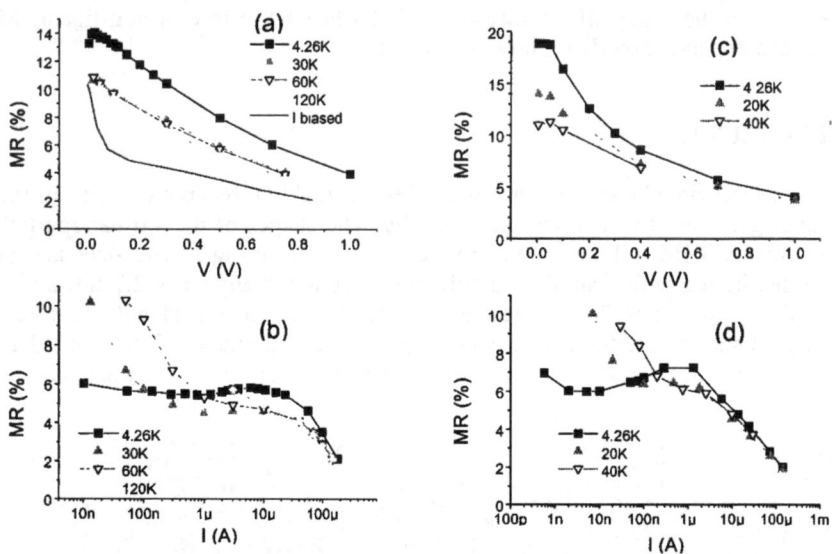

Figure 2 Magnetoresistance versus bias for one layer of Co nanoparticles in the junction; (a)
and (b), and 7 layers of Co; (c) and (d) at various temperatures. The line without symbols in (a)
is I-biased MR at 120K plotted versus saturation voltage.

agrees with Equation (2) above qualitatively. At low temperature, low I-biased MRs are strongly depressed. The difference between I and V-biased MR can be as big as a factor of 5 in other samples. This does not agree with Equation (2). We attribute this to charging effect in the clusters, which give rise to Coulomb blockade effect at low temperatures. The total resistance of the junction is raised while the resistance difference under magnetic field remains.

SUMMARY

We point out that a good magnetic tunnel junction has non-linear current-voltage response, which results in different MR values when it is measured with current bias from with voltage bias. I-biased MR is usually smaller. The ratio between the two cases depends on the physical properties of the junction. When one layer or layers of Co nanoparticles are embedded into the junction, we see a faster decrease of MR with increasing I-bias than with increasing V-bias. There is no significant difference in MR when Co clusters are superparamagnetic of ferromagnetic. At low temperature and low I-bias, MR is strongly depressed. We attribute this to the charging effect in the clusters, which give rise to Coulomb blockade effect at low temperatures.

This work was supported in part by the European Community and the NEDO of Japan.

REFERENCE:

Barnas, J. and Fert, A., 1998, Phys. Rev. Lett. **80**, p1058.
Julliere, M., 1975, Phys. Lett. **54A**, p.225.
Miyazaki, T. and Tezuka, N., 1995, J. Magn. Magn. Mater. **139**, pL231.
Montaigne, F., Nassar, J., Vaurès, A., Nguyen Van Dau, F., Petroff, F., Schuhl, A., Fert, A., 1998, Appl. Phys. Lett. **73**, p2829.
Moodera, J. S., Kinder, L. R., Wong, T. M., and Meservey, R., 1995, Phys. Rev. Lett. **74**, p3273.
Moodera, J. S., Nowak, J., van de Veerdonk, R. J. M., 1998, Phys. Rev. Lett. **80**, p2941.
Nassar, J., Hehn, M., Vaurès, A., Petroff, F., and Fert, A., 1998, Appl. Phys. Lett. **73**, p698.
Schelp, L. F., Fert, A., Fettar, F., Holody, P., Lee, S. F., Maurice, J. L., Petroff, F., and Vaurès, A., 1997, Phys. Rev. **B 56**, pR5747.
Slonczewski, J. C., 1989, Phys. Rev. **B 39**, p6995.
van de Veerdonk, R. J. M., Nowak, J., Meservey, R., Moodera, J. S., and de Jonge, W. J. M., 1997, Appl. Phys. Lett. **71**, p2839.
Zhang, S., Levy, P. M., Marley, A. C., and Parkin, S. S. P., 1997, Phys. Rev. Lett. **79**, p3744.

24 Investigations of GaN Blue Laser Semiconductor

L. W. Tu[1], Y. C. Lee[1], J. W. Tu[1], K. H. Lee[1], K. J. Hsu[1], I. Lo[1]
and K. Y. Hsieh[2]
[1]*Department of Physics and* [2]*Institute of Materials Science and
Engineering, National Sun Yat-Sen University, Kaohsiung, Taiwan
80424, Republic of China*

1.1 INTRODUCTION

In the search of light sources covering the whole visible spectrum, GaN, which has a large direct band gap, is a sure one to study. Bright luminescence in the blue/green range can be seen, and devices in MIS (metal-insulator-semiconductor) structure are obtained in 1971 by Pankove. After more than a decade, breakthroughs emerge in a series of events. In 1983, Yoshida puts down an AlN buffer layer on sapphire before the GaN growth using MBE (molecular beam epitaxy) technique, and gets good results on the properties of GaN films. In 1986, Akasaki uses MOCVD (metal-organic chemical vapor deposition) to grow the AlN buffer layer on sapphire at low temperature, and the GaN film quality improves considerably. Another big leap comes in 1989 when Akasaki successfully activates the Mg dopants using LEEBI (low energy electron beam irradiation) to break up Mg-H bond. Nakamura follows with annealing in an H-free environment to activate the Mg acceptors.

As-grown GaN is mostly n-type due to native defects and/or impurities. Success in p-type doping opens up a whole world in device applications. Not long at all before commercial GaN-based blue/green LEDs with luminous intensity more than 10 cd become a reality. Before the end of 1997, continuous wave GaN-based blue LD (laser diode) at room temperature is achieved. The rapid progress reaches another stage with new ELOG (epitaxial lateral overgrowth, or called LEO for lateral epitaxial overgrowth) structure that yields long lifetime LD devices in accordance with commercial criterion of 10^4 hrs (Nakamura, 1997).

GaN crystal structure can be hexagonal or cubic. Currently, sapphire is the substrate used the most. GaN film grown on sapphire has a hexagonal structure. Table 1.1 lists some properties of GaN semiconductor and the sapphire substrate.

Although commercial products of N-based/sapphire are available on the market already for several years, the mechanism of light emission is under vigorous research. One of the interesting and puzzling phenomena observed on the photoluminescence (PL) spectra of GaN films is the broad luminescence peak

Table 1.1 Properties of GaN and sapphire at 300 K.

	Crystal Structure	Lattice Constant (Å)	Band Gap (eV)	Linear Coefficient of Thermal Expansion (10^{-6}/K)	Thermal Conductivity (W/K·cm)
GaN	Wurtzite	a=3.189 c=5.185	3.4	α_a=5.59 α_c=7.75	1.3
Sapphire	Wurtzite	a=4.758 c=12.991	9.9	α_a=7.5 α_c=8.5	0.5

around 2.3 eV. Many theories and various experimental results can be found in the literature. Among them, deep level models have been proposed by many groups. The physical origins of the deep levels can be ascribed to two kinds, one is due to impurities, and one is due to native defects. In this work, MOCVD grown GaN thin films are dry etched through reactive ion etching (RIE). PL and reflectivity measurements are performed before and after each etching. Detailed analysis on the data shows that the yellow light emitters are confined mainly in the vicinity of the interface of GaN and the sapphire substrate. This result supports the physical origin of the yellow luminescence to be due to native defects.

1.2 EXPERIMENTAL RESULTS AND DISCUSSIONS

The substrates are 2-inch c-sapphire. The GaN film is n-type with a Si doping concentration of 1×10^{18} cm. PL measurements use a 325 nm HeCd laser. Experimental procedures and setup in detail can be found elsewhere (Tu, 1998a, 1998b). All measurements are performed at room temperature.

Figure 1.1 is a typical PL spectrum. A high intensity, sharp peak situated at

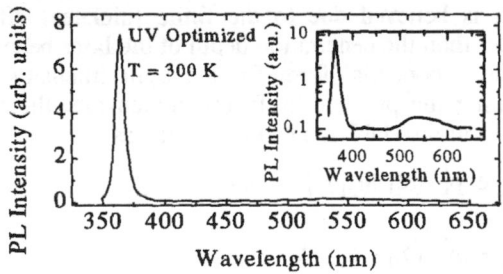

Figure 1.1 Photoluminescence spectrum of GaN/sapphire at room temperature.

364 nm is due to the near-band-gap transition (called the UV peak). While the other broad, less intense peak around 546 nm (see the inset in log scale) is due to

the defect/impurity related transitions (called the yellow peak). While the true origin of the yellow emission is not clear yet, the PL intensity ratio between the UV and the yellow emission is commonly used as an indicator of the quality of a GaN sample. And we found that the PL intensity ratio depends on factors like spectrometer slit width and excitation laser power density (Tu, 1998a).

From SEM measurement of the cross-section of the sample, the thickness of the GaN film is measured as 1.7 μm before the etching. From the interference maxima/minima points of reflectivity spectrum, the GaN film thickness can be calculated as 1.74 μm that is identical with the SEM result.

The relation of the GaN film thickness with the RIE etching time is very linear indicating a good steady etching process and the slope yields an etching rate of 19.2 nm/min. After each 10 min etching, PL is measured as well as reflectivity. The result is shown as in Figure 1.2. Both the UV and the yellow emission

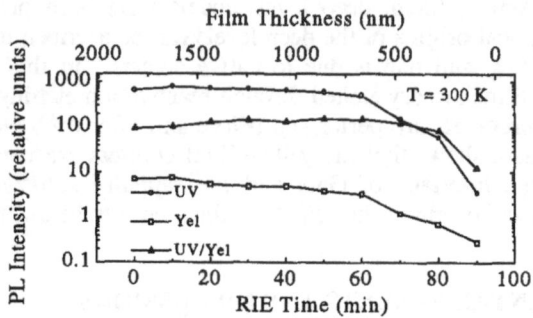

Figure 1.2 Photoluminescence intensity dependency on RIE time.

intensities keep fairly constant with little decease until the thickness of about 700 nm, and drop much faster after that. In the same figure, UV to yellow PL intensity ratio is shown, too, which remains fairly constant until the thickness of about 300 nm. The faster decrease of UV and yellow signal below the thickness of about 700 nm is believed due to the finite thickness effect, i.e., the film thickness is smaller than the penetration depth of the laser beam.

Applying the exponential form for the light intensity absorption in the medium and treating the problem as in one-dimension, the total luminescence emission emitted from the GaN film can be written as

$$I_{UV} = I_0 \eta_1 \int_0^d \exp[-(\alpha_0 + \alpha_1)x]dx, \text{ and} \tag{1.1}$$

$$I_{Yel} = I_0 \eta_2 \int_0^d \exp[-(\alpha_0 + \alpha_2)x]dx, \tag{1.2}$$

where d is the film thickness, I_0 is the laser intensity, α_0, α_1 and α_2 are the absorption coefficients for the laser beam , UV and yellow emission, respectively, and η_1, and η_2 are the generation efficiencies for UV and yellow luminescence, respectively, assuming both are constant. Therefore, knowing that $\alpha_0 > \alpha_1 > \alpha_2$, the ratio of UV to yellow emission, I_{UV}/I_{Yel} , will be decreasing with increasing thickness d, i.e., the ratio will be increasing with increasing RIE etching time, and this is in contradiction to the experimental results. Figure 1.2 shows a fairly constant I_{UV}/I_{Yel} ratio until a film thickness of about 300 nm, and a large drop after that. For a constant η_1 situation, Figure 1.2 shows that η_2 is a function of depth and increases with increasing depth. Based on this analysis, yellow emitter distribution along the film depth is not uniform, and impurities inclusion during the growth should not be the origin for the yellow luminescence. Most yellow emission is from the region near the interface of GaN and sapphire. This is believed to be due to the higher density of defects near the interface region because of the large mismatch of the lattice constant between GaN and the sapphire substrate. Although the possibility that other factors could lower η_1 as well near the interface can not be ruled out. Besides, surface smoothness improves with increasing RIE time as shown in Table 1.2.

Table 1.2 Surface roughness measurements with stylus machine.

RIE time (min)	Ra (nm)	RMS (nm)
0	40.0	45.6
60	34.5	43.2
90	32.7	43.2
90	30.3	40.7

In conclusion, GaN has been investigated with SEM, reflectivity, and PL measurements. Depth profiling of PL measurements on GaN epifilms has been performed using RIE. Film thickness is determined from both the SEM and the reflectivity spectra. An excellent steady etching rate of 19.2 nm/min is established. Film surface smoothness improves with RIE time. Analysis shows that the yellow emitters are mostly confined within the near interface region, and supports the origin of yellow luminescence to be due to native defects instead of impurities.

1.3 REFERENCES

Nakamura, S. and Fasol, G., 1997, *The Blue Laser Diode* (New York: Springer-Verlag).

Tu, L. W., *et al*, 1998a, Spatial Distributions of Near-Band-Gap UV and Yellow Emission on MOCVD Grown GaN Epifilms, *Physical Review B*, **58**, pp. 10696-10699.

Tu, L. W., *et al*, 1998b, Yellow Luminescence Depth Profiling on GaN Epifilms Using Reactive Ion Etching, *Applied Physics Letters*, **71**, pp. 2472-2474.

Part 4

NANOCOMPOSITES

25 Layer-by-Layer Assembly: A Way for Multicomposite Ultrathin Films

Zhiqiang Wang, Liyan Wang, Xi Zhang and Jiacong Shen
Key Lab of Supramolecular Structure and Spectroscopy, Department of Chemistry, Jilin University, Changchun 130023, P. R. China

ABSTRACT

Efforts to achieve layered nanoarchitectures are summarized on basis of layer-by-layer assembly, and functional assemblies of these tailored architectures are discussed. With in-depth study on relationship between the microscopic layered architecture and macroscopic function of supramolecular assemblies, it is anticipated that one could obtain miniature device or machine of high efficiency through integration of the assembling process and device fabrication.

1.1 WHAT IS LAYER-BY-LAYER ASSEMBLY?

Ultrathin assembling films have gained increasingly importance because they allow fabrication of supramolecular assemblies of tailored architecture. Layer-by-layer assembly is a new molecular self-assembling technique based on alternating physisorption of opposite charged polyions at liquid/solid interface.(Decher, 1991) Due to many advantages of this method, such as easiness in fabrication, independence on substrate size and topology, good mechanical and chemical stability of the resulting film, it is recognized that the assembling technique is a powerful and versatile means for assembling multicomposite supramolecular structures with control over the layer composing and thickness.

1.2 EFFORTS TO ACHIEVE LAYERED NANOARCHITECTURES

Layered materials on the basis of macromolecules provide diversities and stability of the resulting supramolecular assemblies. Poly(maleic acid monoester) and bispyridinium bearing biphenyl mesogenic group were used to assemble multilayer film(Zhang, 1996), based on electrostatic attraction between polyion and bication, combined with van der Waals interaction among mesogenic groups of bications. (Figure1A). Bragg diffraction was found in the XRD pattern of the multilayer film, probably indicating the improvement of the interfacial interpenetrating to some extent for these polyeletrolytes and small polyion systems

Fabricating planar molecules of porphyrin or phthalocyanine with unique physichemical properties into supramolecular array remains a great challenge. Multilayer build-up of these planar functional molecules was obtained through alternate deposition of meso-tetra(4-sulfonyl) porphyrin(TPPS$_4$) or copper phthalocyanine tetrasulfonic acid tetrasodium salt (CuTsPc) with cationic bis(pyridinium) salt based on electrostatic attraction(Zhang, 1994). One step further multilayer assemblies of porphyrin and phthalocyanine were prepared by alternating deposition of oppositely charged rigid planar molecules of sodium (phthalocyanine tetra-sulfonate)cobalt Na$_4$[(CoTsPc)] and (tetrakis (N,N,N-trimethyl-4-aniliniumyl)porphyrin) cobalt [(CoTAP)]Br$_4$, which showed efficient charge transfer in-between layers.(Figure 1 B)(Sun, 1996)

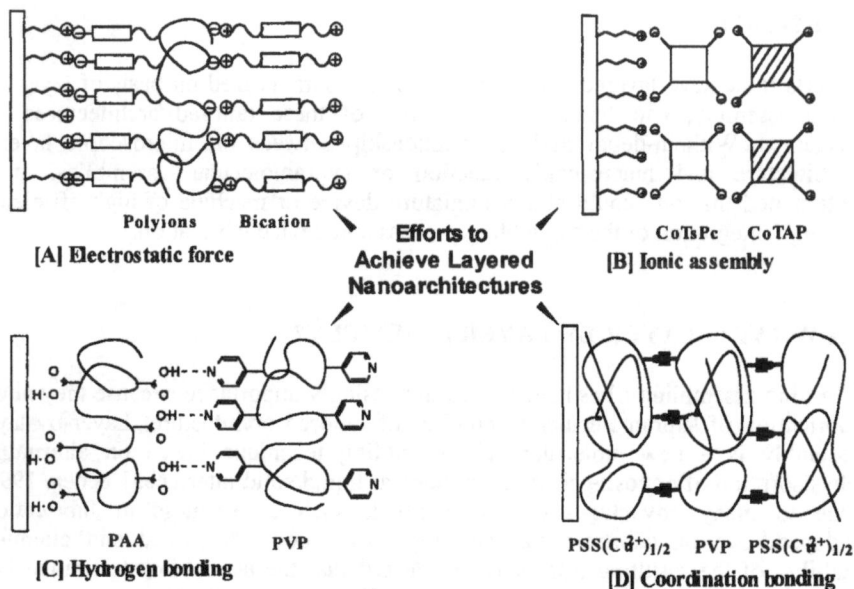

Figure 1 Efforts to achieve layered nanoarchitectures.

Furthermore, the driving force for the alternate assembly need not to be electrostatic. Recently, the multilayer assemblies have been realized by taking advantage of hydrogen bonding of poly (4-vinylpyridine) to poly(acrylic acid) (Figure 1 C)(Wang, 1997). The Kiessig fringes of X-ray diffraction were well resolved and suggested a good and smooth film with constant thickness and low surface roughness. The thickness of the film can be controlled well in nano-scale range by changing the concentration or Mw of the polymer building blocks. Multilayer film can also be fabricated on the basis of coordination bonds, e.g. a self-organizing film via consecutive adsorption poly(copper styrene 4 sulfonate) and poly (4-vinylpyridine)(Xiong, 1998). In this case, the specific coordination between the pyridine and the Cu^{2+} neutralizing the sulfonate group is responsible for build-up of the multilayer (Figure 1 D).

1.3 FUNCTIONAL ASSEMBLIES OF TAILORED ARCHITECTURE

Assembly of enzyme multilayer in a designed way makes it possible not only to mimic biological function but to fabricate bio-reactors. Water-soluble proteins or enzymes can be induced to be charged species similar to polycations or polyanions for preparation of multienzyme assemblies. Protein incorporation in the alternate electrostatic adsorption was realized first with glucose isomerase and a cationic bispyridinium salt(Kong, 1994). Similarly, bienzyme cascade of glucose oxidase and glucoamylase can be fabricated in a designed way(Figure 2 A). Furthermore, no inherent restriction concerning substrate size and topology makes it possible for this assembling technique as a new concept to be used in enzyme immobilization with porous polymer carrier.

Due to the strong adhesion of adjacent layers and independence on substrate topology of this assembling technique, chemically modified electrode in this way may open new horizon for sensors, in particular when combination of flow-injection analysis. For example, CuTsPc-modified gold electrode showed selective potentiometric response to Cu^{2+} ion in the range from 1×10^{-5} to 0.1 mol/L(Figure 2 B)(Sun, 1995). A gold electrode, modified with ultrathin films containing bienzyme of glucose oxidase and glucoamylase, was applied as a bienzyme maltose sensor.

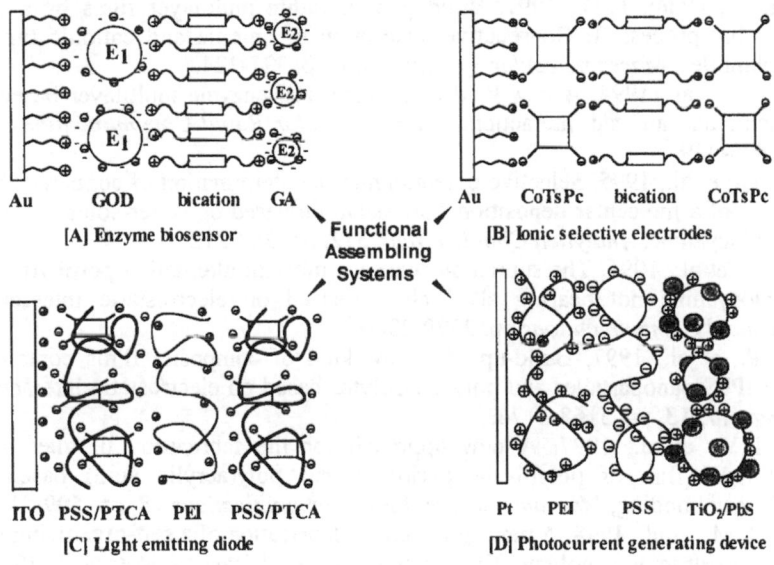

Figure 2 Functional assemblies of tailored architectures.

Self-assembling technique allows a fine-tune of layer thickness as well as interface structure, and thus it could be useful to fabricate organic LED with enhanced efficiency. There exists some dyes and pigments that is not easily assembled directly using alternate deposition technique. Taking the 3,4,9,10-perylenetetracarboxylic acid (PTCA) as an example, it is a very interesting material with good luminescence property. To resolve this problem, we mixed the solution of PTCA with PSS first, and then the mixed solution was used to alternately assemble with poly(ethyleneimine)(PEI). Thus, a stable multilayer assembly was fabricated (Figure 2 C). The electroluminescent device containing 20 layers of PEI/(PSS+PTCA) assembled film as emitting layer was fabricated onto ITO-covered glass electrode as anode, and then covered with aluminium as cathode. A green light was observed with a turn-on voltage of about 4.7 V.

Organizing inorganic nanoparticles into polymer matrix is a way to realize advanced materials of nanocomposite(Sun, 1997). Two approaches have been developed in our lab: (1) preparation of colloidal particles, and then alternate deposition of colloidal particles(e.g., PbI_2 hydrosol, CdS particle, Ag doped ZnS, TiO_2/PbS coupled particles, etc.) with bipolar compounds or polyelectrolytes (Figure 2.D); (2) preparation of self-organizing polymer film, and then in situ formation of heterostructural film by chemical reaction in restricted area. As mentioned previously, $PSS(Cu)_{1/2}$/PVP self-organizing film can be fabricated on the basis of coordination bonding. The heterostructural polymer/Cu_2S nanoparticles film will be form in situ by exposure of this $PSS(Cu)_{1/2}$/PVP multilayer film to H_2S gas.

The work was supported by National Natural Science Foundation of China.

Decher, G., Hong, J. D., 1991, Buildup of ultrathin multilayer films by a self-assembly process: I. Consecutive adsorption of anionic and cationic bipolar amphiphiles. *Macromolecular Symposia*, **46**, pp. 321-324.

Kong, W., et al., 1994, A new kind of immobilized enzyme multilayer based on cationic and anionic interaction, *Macromolecular Rapid Communications*, **15**, pp. 405-409.

Sun, C. Q., et al., 1995, Selective potentionmetric determination of copper(II) ions by use of a molecular deposition film electrode based on water-soluble copper phthalocyanine, *Analytica Chimica Acta*, **312**, pp. 207-212.

Sun, Y.P., et al., 1996, The supramolecular assembly of alternating porphyrin and phthlocyanine not bearing alkyl chains based on electrostatic interaction, *Chemical Communications* pp. 2379-2380.

Sun, Y.P., et al., 1997, Build-up of a new kind of composite films containing TiO2/PbS nanoparticles and polyelectrolytes based on electrostatic interaction, *Langmuir,* **13**, pp. 5168-5174.

Wang, L.Y., et al., 1997, A new approach for the fabrication of alternating multilayer film of poly(4-vinylpyridine) and poly(acrylic acid) based on hydrogen bonding, *Macromolecular Rapid Communications* **18**, pp. 509-514.

Xiong, H. M., et al., 1998, A new approach for fabrication of a self-organizing film of heterostructured polymer/Cu_2S nanoparticles, *Advanced Materials,* **10**, pp. 529-532.

Zhang, X., et al., 1994, Build-up of a new type of ultrathin film of porphyrin and phthalocyanine based on cationic and anionic electrostatic attraction, *Chemical Communications*, pp. 1055-1056.

Zhang, X., et al., 1996, Effects of pH on the supramolecular structure of polymeric molecular deposition films, *Macromolecular Chemistry and Physics*, **197**, pp. 509-515.

26 Sol-gel Preparation and Luminescence Properties of Nano-structured Hybrid Materials Incorporated with Europium Complexes

Bing Yan[1], Hongjie Zhang[1] and Peter A. Tanner[2]
[1]*Key Laboratory of Rare Earth Chemistry and Physics, Changchun Institute of Applied Chemistry, Chinese Academy of Sciences, Changchun 130022, China and* [2]*Department of Biology and Chemistry, City University of Hong Kong, Kowloon, Hong Kong SAR*

ABSTRACT

Microporous silica gel has been prepared by the sol-gel method utilizing the hydrolysis and polycondensation of tetraethylorthosilicate (TEOS). The gel has been doped with the luminescent ternary europium complex $Eu(TTA)_3 \cdot phen$: where HTTA=1-(2-thenoyl)-3,3,3-trifluoroacetone and phen=1,10-phenanthroline. By contrast to the weak f-f electron absorption bands of Eu^{3+}, the complex organic ligand exhibits intense near ultraviolet absorption. Energy transfer from the ligand to Eu^{3+} enables the production of efficient, sharp visible luminescence from this material. Utilizing the polymerization of methyl methacrylate, the inorganic/polymer hybrid material containing $Eu(TTA)_3 \cdot phen$ has also been obtained. SEM micrographs show uniformly dispersed particles in the nanometre range. The characteristic luminescence spectral features of europium ions are present in the emission spectra of the hybrid material doped with $Eu(TTA)_3 \cdot phen$.

1. INTRODUCTION

The sol-gel method has been shown to be a suitable approach for the preparation of novel luminescent materials (Reisfeld, 1997). Photoactive lanthanide-organic coordinated compounds, such as europium and terbium chelates with organic ligands, exhibit intense narrow band emissions via an energy transfer from the ligands to the metal ions under near UV excitation (Yang et al., 1994). Recently, some papers have reported the luminescence behavior of Eu^{3+} and Tb^{3+} complexes with β-diketones (Lowell and Edward, 1993), aromatic carboxylic acids (Zhang et al., 1997) and heterocylic ligands (Jin et al., 1995) in sol-gel derived host materials. Unfortunately, the dopant concentrations of complexes in a silica matrix must be low, and it is difficult to obtain transparent and uniform material. Besides this, the pure inorganic matrix has some disadvantages such as poor mechanical properties that restrict its practical applications.

The incorporation of a polymer into the silica gel matrix forms an inorganic/polymeric hybrid material (Kohjiya et al., 1990; Mascia and Koul, 1994), which can modify the mechanical properties. In the present work, Eu(TTA)·phen was incorporated into a poly(methyl methacrylate) (PMMA) matrix, and the corresponding hybrid material SiO_2/PMMA: Eu(TTA)·phen was also synthesized, so that the compositional, luminescence and thermal properties could be investigated. The ultimate aim is the development of display devices, since the hybrid matrix possesses superior thermal and mechanical properties for incorporation of luminescent europium complexes.

2. EXPERIMENTAL

Methyl methacrylate (MMA) was previously treated with dilute sodium hydroxide solution in order to remove impurities that might hinder polymerization. DMF solutions of $Eu(TTA)_3$·phen (2.0 cm^3) were mixed with MMA (5.3 cm^3), with dopant concentrations (in wt. %) of 0.05, 0.1, 0.5, 1.0, 2.0 and 5.0. Then benzoyl peroxide (BPO) (ca. 0.003 g), as the initiator for the preparation, was added into the mixed solution. The mixture was placed in water bath at 353-363 K for 15 to 30 min. After the pre-polymerization of the monomers, the mixture became viscous, and was placed into a mold or poured along microscope glass slides. The polymerization process was completed in an oven at the 323 K for 24 h.

Microporous silica gel was prepared by the sol-gel method according to hydrolysis and polycondensation process of TEOS (Tamaki et al., 1997). On formation, the silica xerogel was immersed in a DMF solution of MMA containing $Eu(TTA)_3$·phen. The polymerization was first performed at 358 K for 30 min, then at 323 K for 24 h, and finally at 358 K for 1 h to obtain the silica/polymer hybrid material containing $Eu(TTA)_3$·phen.

Thermal analysis was carried out using a Seiko SSC/5200 Thermogravimetric Analyzer. Infra-red absorption spectra were measured on a Bomen MB420 FTIR spectrometer using the KBr pellet technique or a ZnSe ATR accessory. The excitation and emission spectra were recorded using a SLM 4800C Spectrofluorometer equipped with a xenon lamp as the excitation source. UV-visible absorption spectra were measured by a Shimadzu Spectrometer at a resolution of 1 nm. The morphologies of the surface of the blend specimens were observed in a Jeol JSM 820 scanning electron microscope (SEM).

3. RESULTS AND DISCUSSION

Comparison of the TG-DTA curves of Eu(TTA)$_3$·phen, PMMA: Eu(TTA)$_3$·phen and PMMA/SiO$_2$:Eu(TTA)$_3$·phen shows that the temperature for 5% loss in mass of the material, due to decomposition, increases in the order 451 K, 478 K and 590 K. This indicates that the silica gel/polymer hybrid matrix exhibits greater thermal stability than that of pure complex or pure polymer matrix. The silica gel/PMMA hybrid material thus provides a stable structural host matrix for the Eu(TTA)$_3$·phen complex, and is a promising candidate for practical applications.

The maximum absorption peak of Eu(TTA)$_3$·phen in ethanolic solution (λ_{max}) is at 350 nm, and λ_{max} of Eu(TTA)$_3$·phen in silica gel/PMMA is at 346 nm. This band corresponds to an electric dipole transition within the ligand, with the ligand excited state energy being just above that of the $^5D_4$ level of Eu^{3+}. Various nonradiative pathways may populate the luminescent $^5D_0$ level of Eu^{3+} from $^5D_4$. The excitation, energy transfer and luminescence processes of Eu(TTA)$_3$·phen in silica gel/PMMA matrix are thus analogous to those of the neat Eu(TTA)$_3$·phen complex.

The 300 K excitation spectra of Eu(TTA)$_3$·phen in PMMA and in silica gel/PMMA are similar, and show a broad band between 300 nm to 420 nm, with λ_{max} near 370 nm (for example, Fig.1). Figure 2 shows the corresponding emission spectrum, with excitation into the ligand absorption band. The emission transitions centred near 582, 593, 614, 652 and 700 nm correspond to the $^5D_0 \rightarrow$ $^7F_{0,1,2,3,4}$ transitions respectively.

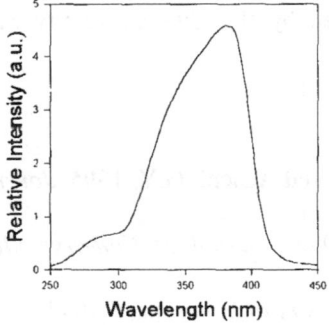
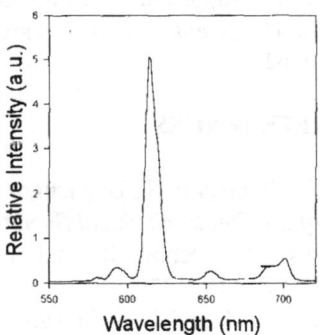

Spectra of 2 wt% Eu(TTA)$_3$.phen in silica gel/PMMA matrix at 300 K
Figure 1 Excitation spectrum   Figure 2 Emission spectrum

The emission intensity of the $^5D_0 \rightarrow {}^7F_0$ transition of Eu^{3+} was investigated, under constant excitation intensity, in both the polymer and hybrid materials as a function of dopant ion concentration. The results of replicate intensity measurements are shown in Figs. 3 and 4, and individual deviations from the overall trends are attributed to slight preparative differences. For Eu(TTA)$_3$·phen in the neat polymer matrix (Fig. 3B), the luminescence intensity reaches the maximum value when the dopant concentration of Eu(TTA)$_3$·phen is about 1 wt.

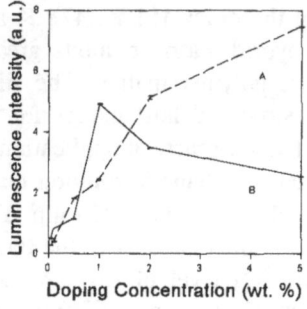
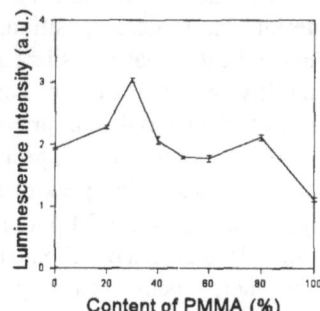

Figure 3 Luminescence intensities of Eu(TTA)$_3$.phen in silica gel/PMMA (A) and PMMA (B)

Figure 4 Luminescence intensities of 0.5 wt % Eu(TTA)$_3$.phen in silica gel/PMMA

%. At higher concentrations, quenching of $^5D_0$ luminescence begins, because although the concentration of luminescent centres increases, the nonradiative relaxation of the $^5D_0$ state *via* several quanta of the C-H stretching modes in the PMMA matrix becomes important. By contrast, the quenching is not evident in the hybrid material up to 5 wt % dopant concentration (Fig. 3B), since the Si-O mode stretching frequencies are about 3 times lower than those of C-H modes. Fig. 4 shows the overall decrease in luminescence intensity in the hybrid material when the proportion of PMMA is increased.

Acknowledgment: This research was supported by the City University SRG 7000762.

4. REFERENCES

Jin, T., Tsutsumi, S., Deguchi, Y., Machida, K.I. and Adachi, G.Y. 1995, *Journal of the Electrochemical Society*, **142**, L195.

Kohjiya, S., Ochiai, K. and Yamashita, S. 1990, *Journal of Non-crystalline Solids*, **119**, 132.

Lowell, R.M. and T. K. Edward T.K. 1993, *Chemistry of Materials*, **5** 1697.

Mascia, L. and Koul, A. 1994, *Journal of Materials Science Letters*, **13**, 641.

Reisfeld, R. 1997, *Journal of Luminescence*, **7**, 72-74.

Tamaki, R., Ahmad, E.Z., Sarwar M.I. and Mark, E. 1997, *Journal of Materials Chemistry*, **7**, 259.

Yang, Y.S., Gong, M.L., Li, Y.Y., Lei, H.Y. and Wu, S.L. 1994, *Journal of Alloys and Compounds*, **207/208**, 112.

Zhang, Y., Wang, M.Q. and Xu, J. 1997, *Materials Science and Engineering*, **B47**, 23.

27 Preparation of Narrowly Distributed Novel Stable and Soluble Polyacetylene Nanoparticles

Chi Wu[1,2], Aizhen Niu[1], Louis M. Leung[3] and T. S. Lam[3]
[1]Department of Chemistry, The Chinese University of Hong Kong, Shatin, N. T., Hong Kong, [2]The Open Laboratory for Bond-Selective Chemistry, Department of Chemical Physics, University of Science and Technology of China, Hefei, Anhui, China and [3]Department of Chemistry, Hong Kong Baptist University, Kowloon, Hong Kong

INTRODUCTION

In this report, we demonstrated that upon heating a diblock copolymer of p-methyl styrene and phenyl vinylsulfoxide (MS-b-PVSO) in tetrahydrofuran, the soluble and flexible PVSO block could be converted into an insoluble and rigid PA block via a chemical reaction, resulting a self-assembly of the copolymer chains into a core-shell nanostructure with the PA blocks as the core and the MS blocks as the shell. The reaction and self-assembly rates could be simply controlled by the reaction temperature. This is an important step towards the processing of intractable polyacetylene as a useful and potential material.

Forty years ago, polyacetylene (PA) as a black powder was first discovered by Natta *et al.*(1984) Twenty years later, Chiang *et al.*(1978) found that the doped PA film had a conductivity close to metal. As a potential conducting and non-linear optical material, polyacetylene has been extensively studied (Nalwa, 1997) in the last two decades. However, polyacetylene is so intractable that its many potential applications have been hindered. It has been a long dream in the field to modify polyacetylene so that it could be processed into different real components for real applications.

On the other hand, extensive studies showed (Vannice, 1984; Dai, 1993; Zhang *et al.*, 1995) that block and graft copolymers could form polymeric micelles in solution if one could make one of the blocks insoluble. It is generally accepted that diblock copolymeric micelles have a core-shell nanostructure with the insoluble block as the core and the soluble block as the swollen shell, resembling small molecule surfactant micelles. Normally, the solvent quality was varied by slowly adding a copolymer solution into a non-solvent for one of the polymer blocks. For water soluble block copolymers, one could also vary the solution temperature to induce the micelle formation if one of the blocks could change from hydrophilic to hydrophobic at higher temperatures (Wu and Qiu, 1998). Noted that the micelle formation is thermodynamically driven and the process is usually so fast in dilute solution that its observation is difficult, if not impossible. To our knowledge, the formation kinetics has not yet been reported.

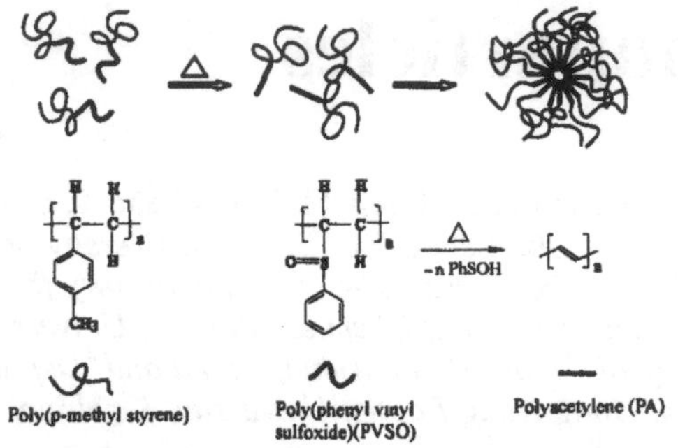

Figure 1. Schematic of chemical reaction-induced self-assembly of poly(4-methyl styrene-*b*-phenylvinylsulfoxide) diblock copolymer chains in solution upon heating.

Recently, Leung *et al.*(1994) reported that poly(phenyl vinyl sulfoxide) could be slowly converted into polyacetylene upon heating. The morphology and the electrical properties of PVSO before and after thermolysis in bulk and its thermolysis kinetics and optical properties in solution have been studied(Leung et al., 1993; Leung et al., 1997). It is expected that if PVSO as one polymer block is connected to another polymer block to form a diblock copolymer, we will be able to observe a chemical reaction induced self-assembly of the copolymer chains because the resulting PA block from the PVSO block is insoluble. Figure 1 shows a schematic of our idea. It should be noted that by properly choosing a reaction temperature in the range 30-80 °C, we are able to control the self-assembly rate and study the self-assembly kinetics in solution.

EXPERIMENT

The MS-*b*-PVSO copolymer (M_w=17,060 g/mol, M_w/M_n=1.12 and n_{MS}:n_{VSO}=117:60) was prepared by a similar anionic polymerization method detailed before. The purification of monomers (MS and PVSO) and solvents were standard. The reaction was initiated by endcapped sec-BuLi. The molar mass and composition of the MS and PVSO blocks and the resulting copolymer were determined by GPC and H^1-NMR, respectively. In this study, the self-assembly of the MS-*b*-PVSO chains in tetrahydrofuran (THF) was monitored by using a modified ALV/SP-150 laser light scattering (LLS) spectrometer with an ALV-5000 time correlator and a HeNe laser with a power of 40 mW at 632.8 nm. The details of LLS can be found elsewhere(Pecora, 1976; Chu, 1991). The Laplace inversion of the measured intensity-intensity time correlation function in dynamic LLS led to the hydrodynamic radius distribution f(R_h).

RESULTS AND DISCUSSION

Figure 2 clearly shows that before the chemical reaction (t = 0), the copolymer chains are narrowly distributed with an average hydrodynamic radius $<R_h>$ located at 3.76 nm. As the reaction proceeded, the solution gradually changed from colorless to dark red and the initial narrow peak in Figure 2 was split into two

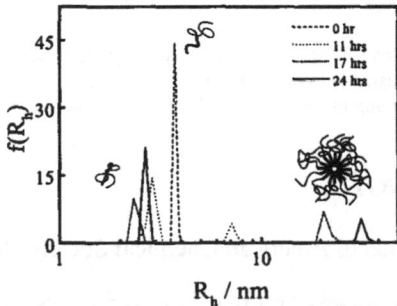

Figure 2. Time dependence of the hydrodynamic radius distribution $f(R_h)$ of poly(4-methyl styrene-*b*-phenyl vinylsulfoxide) in tetrahydrofuran at 55 °C during the reaction, where the initial copolymer concentration was 4.02×10^{-3} g/mL.

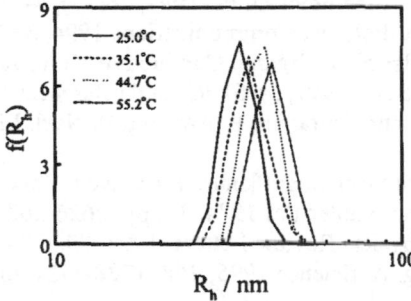

Figure 3. Temperature dependence of the hydrodynamic radius distribution $f(R_h)$ of the resulting polyacetylene core-shell nanoparticles in tetrahydrofuran.

peaks. The appearance of the peak with a larger and increasing size is due to the self-assembly of the resultant MS-*b*-PA chains and the final size of the nanoparticles is in the range 30-60 nm which is expected on the basis of the molar mass and composition of the copolymer chain. The peak around ~2 nm reflects individual copolymer chains (unimers). Initially, we were puzzled by the question why individual MS-*b*-PA chains have a smaller $<R_h>$ than original MS-*b*-PVSO chains. Later, we found that it could be explained by the self-wrapping; namely, the insoluble PA block was wrapped by the soluble MS block so that the interaction between the insoluble PA block and solvent was minimized, which is schematically shown in Figure 2.

We also observed that the nanoparticles were very stable at 55 °C and no color change in the dispersion, indicating that, the PA blocks protected by the MS shell were very stable in the dispersion. Figure 3 shows that the size of the nanoparticles only slightly increases as the solution temperature increases, which is due to the swollen of the soluble MS shell at higher temperatures and further indicates that in the temperature range 25-55 °C, the nanoparticles were very stable in the solution. This is very important in the application of these novel polyacetylene nanoparticles. Our studies also showed that the nanoparticle dispersion could be cast into a thin film or a thin tube or any other desired shapes.

Acknowledgment. The financial supports of this study by the Research Grants Council (RGC) of Hong Kong Government Earmarked Grants 1997/98, (CUHK4181/97P, 2160082, RGC/96-97/03) and The National Distinguished Young Investigator Fund (1996, 29625410) are gratefully acknowledged.

REFERENCES AND NOTES

Chiang, C. K *et al.*, Journal of American Chemical Society 1978, 100, pp1013-1015.

Chu, B. Laser Light Scattering(2nd Ed.); Academic Press, New York, 1991.

Dai, L. M. and White, J. W. Journal Polymer Science Part B. Polymer Physics 1993, 31, pp. 3-15 .

Leung, L. M.; Lam, T. S. Polymer Preprint (Polymer Chemistry Division) 1997, 38(1), pp. 156-157.

Leung, L. M.; Tan, K. H. Macromolecules, 1993, 26(17), 4426-4436.

Leung, L. M.; Tan, K. H. Polymer Communications 1994, 35(7), 1556-1560.

Nalwa, H. S. (Ed.) Handbook of Organic Conductive Molecules and Polymers; vol.2 John Wiley & Sons. 1997; pp 61-95 and references therein.

Natta, G.; Mazzanti, G.; and Corradini, P. Atti Accad. Nazl. Lincei. Rend. Cl. Sci. 1958, 25, pp. 3-12.

Pecora, R. Dynamic Light Scattering; Plenum Press, New York, 1976.

Vannice, F. L. *et al.*, Macromolecules 1984, 17, pp. 2626-2629.

Wu, C. and Qiu X. P. Physical Review Letters 1998, 79, 620-622.

Zhang, L. and Eisenberg, A. Science 1995, 268, 1728-1731 and the references therein.

28 Liquid Crystalline Behaviour of Polystyrene/Clay Nanocomposite[1]

Yongmei Ma and Shimin Zhang
State Key Laboratory of Engineering Plastics, Institute of Chemistry, Chinese Academy of Sciences, Beijing 100080, P. R. China

ABSTRACT

Fully exfoliated polystyrene/clay nanocomposite (PSCN) has been prepared by intercalation polymerization of styrene in the presence of organo-montmorillonite. The injection-molded PSCN sample (IPSCN) exhibits liquid crystalline behavior: XRD patterns revealed a sharp diffraction peak at $2\theta = 1.7°$ and strong birefringence was observed under POM for the IPSCN sample. The nanocomposite shows unique thermal characteristic from DSC measurement. The relaxation of birefringence in IPSCN has also been studied.

INTRODUCTION

Polymer/clay organic/inorganic hybrid nanocomposites (PCN) have received significant attention in recent years, because they exhibit unexpected properties not sharing by their conventional counterparts. One of the promising routes to prepare polymer/clay nanocomposite is via intercalation polymerization process. In this process, the clay used is mainly montmorillonite (MMT). The layer

---

[1] This work was partially supported by the NSFC and China National 863 Programme.

structure of MMT consists of two silica tetrahedral and an alumina octahedral sheets, stacking of the layers of ca. 1 nm thickness by weak dipolar forces leading to galleries between the layers (ca. 1 nm). The galleries are occupied by cations, such as Na^+, Ca^{2+}, Mg^{2+}, that are easily replaced to form organo-clay by alkylammonium ion exchange reaction. The organo-clay can react physically or chemically with monomers (Giannelies, 1996; Okada and Usuki, 1995). During polymerization, the clay can be exfoliated into their nanoscale building blocks and uniformly disperse in the polymer matrices to form exfoliated PCN.

Up to now, most of liquid crystalline polymers is composed of mesogenic groups in the main or side chains and flexible spacer segments (Plate *et al*, 1991). Under certain conditions, however, some polymers not containing rigid liquid-crystalline segments but meeting the structural characteristics also possesses liquid crystalline phase. For instance, Bassett (1981) and Rastogi *et al* (1991) found respectively that polyethylene and poly(4-methyl-pentene-1) generate liquid crystalline phase under certain pressure range. By analogue of thermotropic and lyotropic, such type of liquid crystal is named barrotropic. Recently, Qi and Zhang *et al* (1998) firstly reported novel liquid crystalline materials of polymer/clay nanocomposites (PSCN), which does not contain any mesogenic group. In this present paper, we discuss the structure feature of this nanocomposite from the experimental results obtained by POM, DSC, TEM and XRD.

EXPERIMENTAL

The synthesis of PSCN were previously reported by Qi *et al* (1998). In brief, the polystyrene/clay nanocomposite was prepared by emulsion polymerization of styrene in the presence of organo-montmorillonite (organo-MMT) obtained by intercalation with cetyltrimethyl ammonium bromide. The injection-molded samples (IPSCN) is prepared with CS-183 Mini-Max molder. XRD patterns were obtained with Rigaku D/Max-2400 X-ray diffracter, Cu Kα, 40 KV, 40 mA. A Hitachi 800 transmission electron microscope (TEM) was used to observe the microstructure of the nanocomposite slice at an acceleration voltage of 100KV. The IPSCN samples were photographed using an Olympus BHSP polarization optical microscope (POM) with a Mettler FP-52 hot stage. Thermal properties were characterized with Perkin-Elmer DSC7 differential scanning calorimeter at heating rate of 20°/min.

RESULTS AND DISCUSSION

Typical XRD patterns of PSCN are shown in Fig. 1. Organo-MMT (curve a) exhibits evident diffraction at $2\theta = 4.5°$, while PSCN containing 5% MMT (curve b) gives no peak attributable to the (001) plane of MMT until $2\theta = 0.5°$. This suggests that the MMT layer were fully exfoliated and randomly dispersed in the polymer matrices at the nanometer level.

In order to get the direct evidence of the nanostructure of PSCN, TEM was employed to observe the slice of the nanocomposites and the photograph is shown in Fig. 2. Imagine analysis indicates that MMT is uniformly dispersed as lamellas

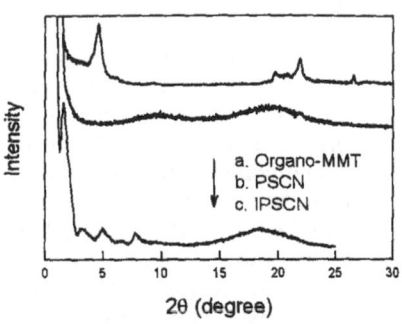

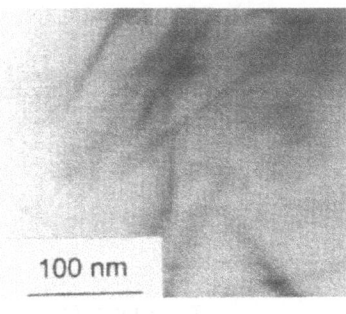

Fig. 1 XRD patterns of organo-MMT and PSCN.   Fig. 2 TEM photograph of PSCN.

of 2~8 nm in the polymer matrix.

PSCN sample exhibited excellent ability to orientation during the process of injection-molded. The XRD patterns reveal the difference between these two sample, while the unprocessed PSCN (Fig 1, curve b) shows only a disperse shoulder at the wild-angle zone ($2\theta = 19.5°$) which is the characteristic diffraction of amorphous PS, the IPSCN (Fig 1, curve c) emerges a sharp diffraction peak at a small angle ($2\theta = 1.7°$) and a disperse shoulder at $2\theta = 18°$. The sharp diffraction at such a small angle indicates that the nanocomposite possesses long-term-ordered layered structure like common liquid crystalline polymers. In other words, the structure of IPSCN falls within the categories of liquid crystal.

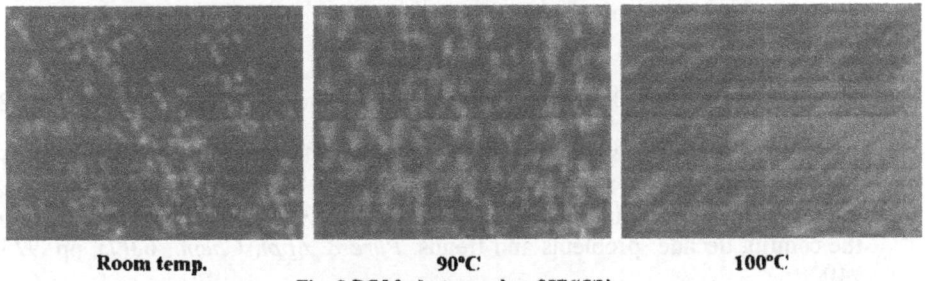

Room temp.   90°C   100°C
Fig. 3 POM photographs of IPSCN.

Fig. 3 shows the POM photographs taken from IPSCN. At the room temperature, the nanocomposite visualizes clear birefringence like a liquid crystalline polymer and the liquid crystalline phase is evidently oriented. With increasing temperature, the textural structure gradually relaxes and completely disappears after the glass transition of the polymer. In contrast of IPSCN, the PSCN does not show any visible liquid crystalline characteristic.

This nanocomposite has shown unique thermal characteristic. The DSC scans show that IPSCN does not possess other evident phase transition near the Tg although there is a dual-inflexion transition when the sample is changed from glass state to elastomer state. Similar phenomenon was found by Zhou (1998) in a conventional liquid crystalline polymer, in that case the liquid crystalline characteristic disappears after the Tg and no evident transition was then detected

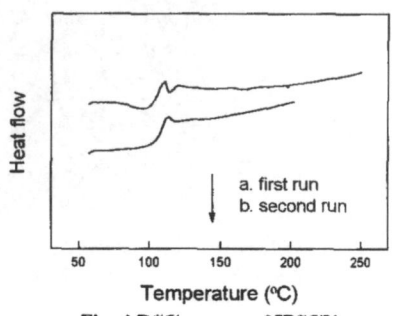

a. first run
b. second run

Temperature (°C)

Fig. 4 DSC curves of IPSCN.

by DSC. When the IPSCN sample was first heated, the DSC curve appeared dual-inflexions near the Tg of polystyrene. While the sample was secondly heated, there was only one inflexion in the DSC curve. This phenomenon implies that there is mesostate at injection-molded sample and the mesostate is not stable at high temperature. We consider this mesostate corresponds to the liquid crystalline morphology, which is agreed with the POM observation discussed above.

Bulk PS is an amorphous polymer, which contains no liquid crystalline segments. When PS was formed nanocomposite with organo-montmorillonite, the clay layer exfoliated and dispersed into the polymer matrices, the interaction between polymer chains and the nanoscale clay layers is strong due to the nanometer effect. The result might cause the nanocomposite to meet the structural characteristics of liquid crystalline polymers. When the nanocomposites are processed, the field of the shearing force farther induces the liquid crystalline behaviour of the samples. Because clay layers restrict the movement of the polymeric matrix. The liquid crystal phenomenon is still visible even though the force field is removed from the processed nanocomposite samples.

REFERENCES

Giannelies, E.P., 1996, Polymer layered silicate nanocomposites. *Adv. Mater.*, **8**, pp. 29-35.

Okada, A. and Usuki, A., 1995, The chemistry of polymer-clay hybrid. *Mater. Sci. Eng. C3*, pp. 109-112.

Plate, N.A., Kulichikhin, V.G. and Talroze, R.V., 1991, Mesophase polymers in the coming decade: problems and trends. *Pure & Appl. Chem.*, **63**(7), pp. 925-940.

Bassett, D.C., 1981, *Principles of Polymer Morphology*, (Cambridge: Cambridge University Press), p 169.

Rastogi, S., Newman, M. and Keller, A., 1991, Pressure-induced amorphization and disordering on cooling in a crystalline polymer. *Nature*, **353**, pp. 55-57.

Qi, Z.N., Zhang, S.F., Ma, Y.M., Chen, G.M., Zhang S.M. and Wang, F.S., 1998, Liquid crystalline behavior of polymer/clay nanocomposite. In *Proceedings of the 6th National Symposium on Liquid Crystalline Polymers* (Xiangtan, China, 1-6 October, 1998), Supplement.

Qi, Z.N., Ma, Y.M., Wang F.S. and Chen, G.M., 1998, A polystyrene/clay nanocomposite and its preparation. *China Patent* 98103038.6

Zhou, Q.F., 1998, private communication.

29 Luminescent Properties of Nano-structured Monolith Containing Rare Earth Complexes

Xiang Ling Ji[1], Bin Li[2], Hong Jie Zhang[2], Xia Bin Jing[1] and Bing Zheng Jiang[1]

[1]*Polymer Physics Laboratory and* [2]*Laboratory of Rare-Earth Chemistry and Physics, Changchun Institute of Applied Chemistry, Chinese Academy of Sciences, Changchun 130022, P. R. China*

ABSTRACT

Transparent polymer/rare-earth complex (Eu(TTA)$_3$Phen, Tb(Sal)$_3$) nano-structured monoliths were prepared via sol-gel technique. It could be observed by transmission electron microscope that the fluorescent particles are distributed in the matrix and their size is in the range of 1 to 100nm controlled by reaction conditions. The matrix is composed of organic-inorganic interpenetrating networks.

The fluorescence emission spectra of samples are still similar to corresponding powdered rare-earth complexes. And the half-widths of the strongest bands are less than 10nm. This indicates that the monolith exhibits high fluorescent intensity and color purity. The fluorescence lifetimes of samples are longer than the original rare-earth complexes, respectively.

Samples irradiated with a UV lamp(365nm) are still transparent but become bright red and green in color due to fluorescence of Eu(III) and Tb(III) complexes.

INTRODUCTION

Rare earth compounds often exhibit efficiently luminescent, magnetic and catalytic behaviors[1]. Its emission provides good efficiency and high colormetric purity by the 4f electron transition. Usually, the powdered rare earth compounds are prepared through sintering at high temperature. However, it is difficult to incorporate rare earth compounds into inorganic or organic matrix to achieve a uniform dispersion at the nano-scale. Fortunately, rare earth complexes with organic ligands have shown good solubility in common organic solvents. They can absorb ultraviolet ray in the ligands and produce red-emitting (Eu(III)) and green-emitting (Tb(III)) with good efficiency via the subsequent intramolecular energy transfer from ligands to rare earth ions. Recently the incorporation of inorganic ions into silica gels and glasses via sol-gel technique is of interest for a variety of technological applications, including optical devices such as fiber amplifiers and solid-state lasers[2-4]. Of particular interest are rare-earth-ion-based systems[5-7]. Sol-gel technology provides an attractive alternate route to the preparation of such kind of inorganic-organic composites since it offers a degree of control over the microstructure and composition of the host matrix, and provides an opportunity to prepare new materials. The most common approach involves dissolving a monomer, an oligomer or a polymer in a TEOS(tetraethyl orthosilicate) solution and then allowing hydrolysis and polycondensation to form an inorganic network. Under appropriate conditions, the polymer remains uniformly embedded within the inorganic gel after aging and drying. Polymers   improve the flexibility and facilitate the processing of composites.

EXPERIMENTAL

Materials

Rare Earth Complexes
Eu(TTA)$_3$Phen (HTTA: trifluorothenoyl-acetone, Phen: 1, 10- phenanthroline) and Tb(Sal)$_3$ (Hsal: salicylic acid) were provided by Laboratory of Rare Earth Chemistry and Physics. They are characterized by IR, UV, TGA and elemental analysis and are proved to be the destination products.

Monoliths
In Table 1, various compositions of reactants for the composites are listed. All the reactants weighed in the given proportions were dissolved into DMF in a beaker to get a colorless and transparent solution. When the reaction reached the gel point, the beaker was put into a desiccator for removal of the water and solvents. The drying process was very slow and lasted for 2-3 months.   Finally, the heat treatment at 70-100°C for 2 days is necessary.

Measurements

Transmission Electron Microscopy
The sample was ultramicrotomed to ultrathin film with an ultramicrotoming machine, model LKB 8800-III, made in Sweden. The prepared specimens were observed with a JEM-2010 transmission electron microscope (JEOL Co. , Japan) operated at a voltage of 200KV.

Fluorescence Spectra

The fluorescence spectrum of the specimen was measured with Spex FL-2T2 spectrophotometer with exciting wavelength of 350 nm, emitting wavelength of 612 nm, and the width of the slit is 0.1 mm.

Fluorescence lifetime was measured through a Spex 1934D phosphorescence spectrophotometer with exciting wavelength of 350 nm.

Table 1. Compositions of reactant systems

Sample	TEOS (g)	HEMA (g)	H_2O (g)	pH	DMF (g)	BPO (mg)	$Eu(TTA)_3Phen$ (mg)	$Tb(Sal)_3$ (mg)
Eu-A	2.0	0	0.9	5	3.0	0	10	
Eu-B	1.6	0.4	0.72	5	2.5	4	10	
Eu-C	1.2	0.8	0.54	5	2.0	6	10	
Eu-D	0.8	1.2	0.36	5	1.5	12	10	
Tb-A	2.0	0	0.9	5	3.0	0		10
Tb-B	1.6	0.4	0.72	5	2.5	4		10
Tb-C	1.2	0.8	0.54	5	2.0	6		10
Tb-D	0.8	1.2	0.36	5	1.5	12		10

TEOS: tetraethyl orthosilicate; HEMA: hydroxyethyl methacrylate; DMF: N,N-dimethylformamide; BPO: benzoyl peroxide

RESULTS AND DISCUSSION

The transmission electron micrographs of the ultramicrotomed sample were shown in Fig. 1(Samples Eu-B, Eu-D). The dark domains are fluorescent particles due to higher electron density of Eu(III) compound. It could be seen that they are distributed in the composite and their size is about 60-100nm for Eu-B and 3-6nm for Eu-D. Controlling the reaction conditions such as temperature, concentration, time, etc., the size of the rare earth complex dispersed in the matrix is in the range of 1-100nm. Regarding the matrix, it is composed of both silica network and PHEMA chain, i.e. inorganic-organic interpenetrating network.

The fluorescence spectra of samples Eu(III) series and Tb(III) series were shown in Fig.2 and Fig.3. The spectra of samples are still similar to corresponding powdered rare-earth complexes, i.e. Eu(III) $^5D_0 \rightarrow {}^7F_i$ (i=0,1,2,3,4), where $^5D_0 \rightarrow {}^7F_2$ is the strongest transition located at 612nm, Tb(III) $^5D_4 \rightarrow {}^7F_j$ (j=6,5,4,3), where $^5D_4 \rightarrow {}^7F_5$ is the strongest transition located at 545nm. And the half-widths of the strongest bands are less than 10nm. The measurement was carried out with a very narrow width of slit , 0.1 mm. These indicate that the monolith exhibits high fluorescent intensity and color purity. Because the radiative transition of rare earth ions belongs to f-f transition, electrons in $4f^n$ shielded by both 6s and 6p shells are less influenced by external circumstances. Thus, the interpenetrating network do not destroy the energy transfer from ligands to rare earth ions.

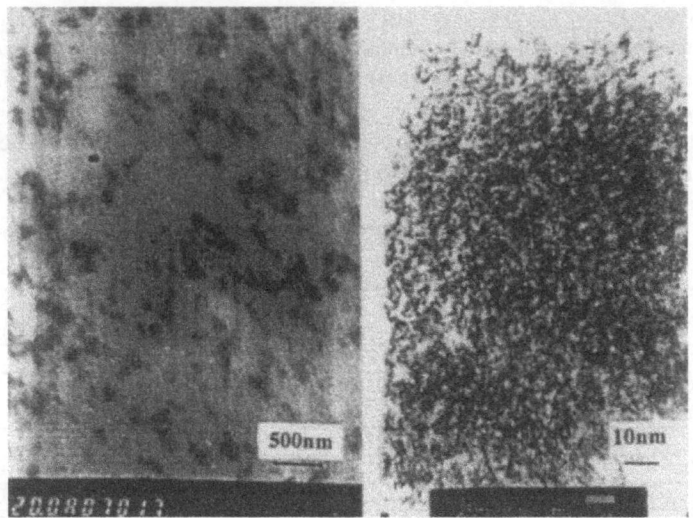

Figure 1 Transmission electron micrographs of the ultramicrotomed
sample Eu-B (left) and Eu-D(right)

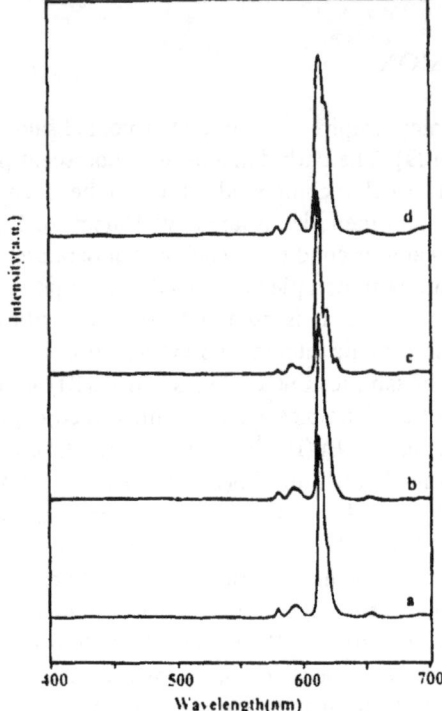

Figure 2 Fluorescence spectra of (a) pure $Eu(TTA)_3$ Phen
(b)Eu-A (c) Eu-B (d) Eu-D

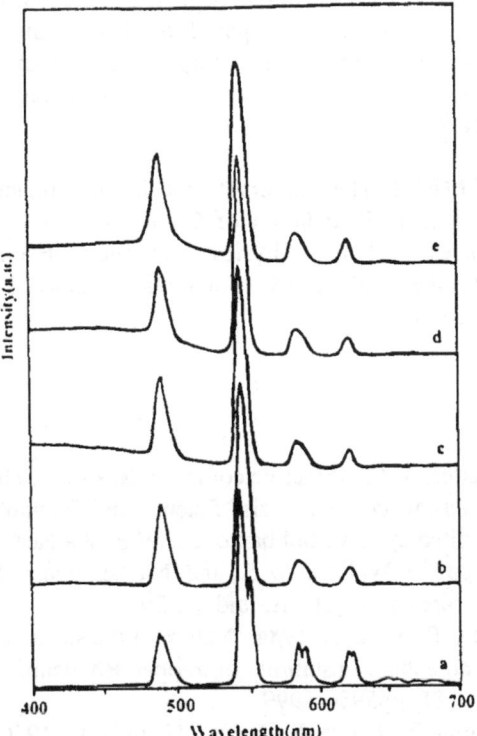

Figure 3 Fluorescence spectra of (a) pure Tb(sal)3
(b) Tb-A (c)Tb-B (d)Tb-C (e)Tb-D

The fluorescent lifetimes of the samples were listed in Table 2. The lifetimes of monoliths are longer than the original rare-earth complexes, respectively. Because the complex in the networks was restricted by the dense solid matrix, the molecular motion is too much limited and stretching vibration of bonds weakened, which decrease the non-radiative transition, thus every monolith has a longer lifetime.

Table 2. The fluorescent lifetime

sample	$Eu(TTA)_3Phen$	Eu-D
$\tau(ms)$	0.82	1.42

CONCLUSIONS

The nano-structured monoliths containing rare earth complexes (Eu(III), Tb(III)) could be prepared with the sol-gel technique by using TEOS and a monomer as starting materials; the rare earth complex dispersed in the matrix is in the range of 1-100 nm in size, the monolith prepared in this manner is crack-free, the fluorescence of nanocomposites exhibits high intensity and colour purity, the fluorescence lifetimes of monoliths are longer than the original rare-earth complexes, respectively.

ACKNOWLEDGEMENT The authors thank for the financial supports from National Natural Scientific Foundation of China, Key Foundation of Chinese Academy of Sciences, Polymer Physics Laboratory Foundation, Rare-earth Chemistry and Physics Laboratory Foundation, Science and Technology Committee of Jilin Province.

REFERENCES

1. Wallace, W. E., Elatter, A. Intermetallic compounds: surface chemistry, hydrogen absorption and heterogeneous catalysis, Science and Technology of Rare Earth Materials, Edited by E. C. Subbarao and W.E. Wallace, Academic Press, New York, 1980, pp329; Wallace, W.E., and Narasimhan, K. S. V. L., Magnetic properties of 2:17 rare earth systems, ibid, pp 393
2. Avnir, D., Lery, D., Reisfeld, R. 1984, Nature of the silica cage as reflected by spectral changes and enhanced stability of trapped Rhodamine 6G, Journal of Physical Chemistry, 88, pp5956-5999
3. Knobbe, E. T., Dunn, B., Fuqua, P. D. And Nishida, F., 1990, Laser behavior and photostability characteristics of organic dye doped silicate gel materials, Applied Optics, 29, pp2729-2733
4. Berry, A. J. And King, T. A., 1989, Characterisation of doped sol-gel derived silica host for use in turnable glass lasers, Journal of Physics D: Applied Physics, 22, pp1419-1422
5. Matthews, L. R. And Knobbe, E. T., 1993, Luminescence behavior of europium complexes in sol-gel derived host materials, Chemistry of Materials, 5, pp1697-1700
6. Jin, T., Tsutsumi, S., Deguchi, Y., Machida, K. and Adachi, G., 1995, Luminescence property of the terbium bipyridyl complex incorporated in silica matrix by a sol-gel method, Journal of Electrochemical Society, 142, L195
7. Jin, T., Inoue, S., Tsutsumi, S., Machida, K. and Adachi, G., 1998, Luminescence properties of lanthanide complexes incorporated into sol-gel derived inorganic-organic composite materials, Journal of Non-Crystalline Solids. 223, pp123-132

30 Alternating LB Film of Ferric Oxide Nanoparticles-Copper Phthalocyanine Derivative and its Gas Sensitivity

L. H. Huo[1,2], L. X. Cao[1], L. Y. Wang[1], H. N. Cui[1], S. Q. Xi[1],
Y. Q. Wu[2], M. X. Huo[2] and J. Q. Zhao[2]
[1]Changchun Institute of Applied Chemistry, Chinese Academy of
Sciences, Changchun 130022 and [2]College of Chemistry and Chemical
Engineering, Heilongjiang University, Harbin 150080, P. R. China

ABSTRACT

Nanoparticulate ferric oxide – tris - (2,4-di-t-amylphenoxy) - (8-quinolinolyl) copper phthalocyanine Langmuir-Blodgett Z-type multilayers were obtained by using monodisperse nanoparticle ferric oxide hydrosol as the subphase. XPS data reveal that the nanoparticle ferric oxide exist as α-Fe_2O_3 phase in the films. Transition electron microscopic (TEM) image of the alternating monolayer shows that the film was highly covered by the copper phthalocyanine derivative and the nanoparticles were arranged rather closely. IR and visible spectra all give the results that the nanoparticles were deposited onto the substrate with the copper phthalocyanine derivative. The gas-sensing measurements show that the alternating LB film had very fast response-recovery characteristic to 2 ppm C_2H_5OH gas, and also sensitive to larger than 200 ppm NH_3.

This work was supported by National Natural Science Foundation of China and Youth Foundation of Heilongjiang Province.

INTRODUCTION

Inorganic oxide nanoparticles organized assembly can be prepared by compositing with organic compounds which have good film-forming ability(Li L. S. *et al.*,1997, Peng X. G. *et al.*,1992, Zhang J. *et al.*,1997). And new properties can be obtained when appropriate functional materials are chosen as the composite materials(Zhang J. *et al.*,1997). The research on inorganic-organic composite material, especially the ordered inorganic nanoparticles-functional organic molecules composite thin or ultrathin films, is being attracted much attention in recent years. Langmuir-Blodgett technique is one of the most generally employed methods to prepare this kind of ordered films. In this paper, alternating LB film of ferric oxide nanoparticles-copper phthalocyanine derivative and its gas sensitivity were investigated.

EXPERIMENTAL DETAILS

Ferric oxide hydrosol used as subphase was prepared as reference(Peng X. G. *et al.*,1992). The concentration of Fe^{3+} was 3.0×10^{-3} mol dm^{-3} determined by TJA POEMS ICP-AES spectroscopy. Tris-(2,4-di-t-amylphenoxy)-(8-quinolinolyl) copper phthalocyanine was synthesized according to paper(Zhang Y., Chen W. Q., Shen Q.,1993) and purified by column chromatography. The spreading solvent was chloroform of A.R. grade, made in China. The concentration of copper phthalocyanine derivative-chloroform solution was 1.0×10^{-4} mol dm^{-3}. The alternating LB films were deposited at constant surface pressure(30mN m^{-1}) at room temperature by using KSV-5000 system. The dipping speed was 5mm min^{-1}. The transfer ratios of Z-type LB film were in the range of 0.95-1.05.

The composition of the LB film were measured using VG ESCALAB MK-II X-ray photoelectron spectrometer with Al Kα radiation(hν=1486.6eV). TEM studies of monolayer were carried out with JEOL-2100 electron microscope. IR and visible spectra were obtained from Bio-Rad FTS-135 and Perkin-Elmer Lambda 9 spectrophotometers, respectively.

The electrical resistance of LB films under various gas concentrations at atmospheric temperature was measured with a set of apparatus equipped by ourselves. Air was used as the diluted gas. The sensitivity is defined as the ratio Ra/Rg, where Ra and Rg are the resistance of LB film measured in air and in air containing the tested gas, respectively.

RESULTS AND DISCUSSION

XPS Result and TEM Image

XPS is a useful means to determine the structure of LB films. The *composition of*

15-layer Fe_2O_3-CuPc alternating LB film deposited on CaF_2 substrate was measured by XPS(see Fig. 1). It reveals that the alternating LB film is composed of copper phthalocyanine derivative and ferric oxide. And the binding energy of $Fe_{2P3/2}$(711.1 eV) implies that the ferric oxide nanoparticles are in α- Fe_2O_3 phase(Peng X G. *et al.*,1992).

TEM image of Fe_2O_3-CuPc derivative monolayer is given in Fig.2. It can be seen that the nanoparticles are arranged equally and closer to each other. After keeping the sample in air for about one month, there is no obvious change observed in TEM image. This result suggests that the nanoparticles can be well restrained to grow or segregate by the surface film-forming molecules.

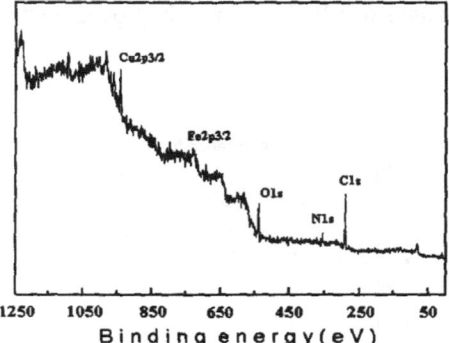

Figure 1 XPS pattern of alternating film   Figure 2 TEM image of monolayer

IR Spectra

The IR spectrum of the 15-layer alternating LB film is similar with that of pure copper phthalocyanine derivative LB film(Ding H. M. et al, 1997), except the intensity change of 1263 and 1230 cm^{-1} bands. For the ferric oxide nanoparticle doping, the intensity of 1230 cm^{-1} was weakened while the 1263 cm^{-1} band was changed from a weak boundary band to a distinctly sharp one. It is due to the interaction between the oxygen of side chain in copper phthalocyanine derivative and ferric oxide nanoparticles.

Visible Spectra

The visible spectrum of copper phthalocyanine derivative LB film is also influenced by incorporation the film-formation of the ferric oxide nanoparticles. The unique absorption band at 615nm(the spectra is omitted) is the dimerric absorbance of Q-band in phthalocyanine ring, which is the same as that in pure copper phthalocyanine derivative LB film(Snow A. W. and Javis N. L., 1984). The band is shifted about 3nm to lower wave. It is resulted from the electron density decreased owing to the interaction between the nanoparticles and phthalocyanine macroring.

Gas Sensitivity

The gas-sensing measurements indicated that the 31-layer alternating film was very sensitive to 2 ppm of C_2H_5OH gas and 200 ppm of NH_3 gas at room temperature. The gas concentrations dependence of the gas sensitivity are given in Fig. 3. From Fig. 3, we can see that good linearity relationships between the sensitivity and gas concentration (C_2H_5OH---2~8 ppm, NH_3---200~2000 ppm) were obtained. Therefore the alternating film can be used as the C_2H_5OH sensor in the range of 2~8 ppm for the fast response-recovery characteristics. The response and recovery times are 15 seconds and 3 minutes, respectively. This result indicates that the doping nanoparticles influence the gas sensitivity of pure copper phthalocyanine derivative LB film(the pure copper phthalocyanine derivative LB film was only sensitive to NH_3 gas). And the alternating LB film is more stable than pure one in air. All the results suggest that the performance of LB film can be improved by the appropriate composite of some materials.

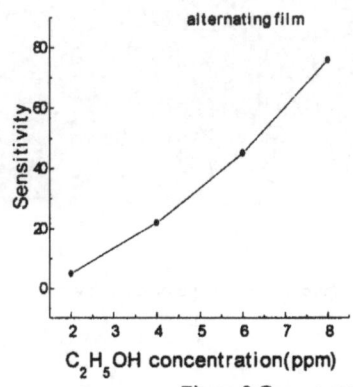 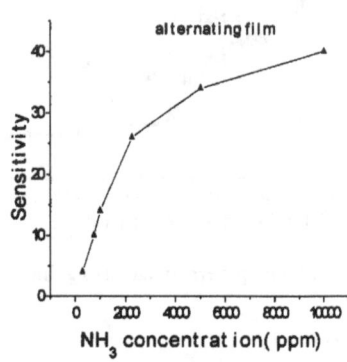

Figure 3 Gas concentration dependence of the gas sensitivity

References:

Ding H. M., Zhang Y., Chen W. Q.,et al, 1997, Structural characterization of copper-phthalo-cyanine thin solid films by FTIR spectroscopy, Spectroscopy and spectral analysis, 17(2), pp. 73-76

Li L. S., Hui Z., Chen Y. M., et al., 1997, Preparation and organized assembly of nanoparticulate TiO_2-stearate alternating Langmuir-Blodgett films, Journal of Colloid and Interface Science, 192, pp. 275-280

Peng X. G., Zhang Y., Yang J., et al., 1992, Formation of nanoparticulate Fe_2O_3–stearate multilayer through the Langmuir-Blodgett method, Journal of Physical Chemistry, 96, pp. 3412-3415

Snow A. W., and Javis N. L., 1984, Molecular association and monolayer formation of soluble phthalocyanine compounds, Journal of American Chemical Society, 106, pp. 4706-4711

Zhang J., Wang D. J., Chen Y. M., et al., 1997, A new type of organic-inorganic multilayer: fabrication and photoelectric properties, Thin Solid Films, 300, pp. 208-212

Zhang Y., Chen W. Q., Shen Q., 1993, Synthesis of copper (II) phthalocyanine derivatives and their LB films, Chemical journal of chinese universities, 14 (11), pp. 1483-1486

31 Photoconductive TiOPc Nanoparticles Embedded in Polymer Prepared from LPDR

Hong-Zheng Chen, Chao Pan and Mang Wang
Department of Polymer Science and Engineering, Zhejiang University, Hangzhou 310027, P. R. China and State Key Lab of Silicon Materials, Hangzhou 310027, P. R. China

INTRODUCTION

Recently, much more attention has been paid to the nanometer materials because of their unique characteristics. However, most of the materials studied are inorganic compounds, such as metals, oxides and nonmetals, which are formed by metal bonds or covalent bonds, and very few reports are about the preparations of the organic nanometer particles[1]. However, Koyman *et al.* found that the photoconductivity of the organic azo pigment depended on the size of the crystal[2]. Nanoscale size effects on photoconductivity of conjugated polymers are also observed by X.Zhang et al.[3]

Nanoparticles in organic polymers or inorganic glasses are of a particular interest both in fundamental viewpoint and application for photoelectronic devices[4]. In the paper, oxotitanium phthalocyanine (TiOPc), as an organic photoconductive pigment, was fabricated into nanoscale particles embedded in polymer employing the means of the liquid phase direct reprecipitation (LPDR).The optimal experimental conditions for preparing TiOPc nanometer particles were studied. The photoconductivities of TiOPc nanometer

TiOPc

particles in either double-layered photoreceptor (P/R) or single-layered P/R were investigated as well.

EXPERIMENTAL PART

Materials

TiOPc was synthesized and purified according to the literature [5]. Casein, polycarbonate (PC), α -naphthalic hydrazone(α -NH) , polyamide (PA), polyvinyl butyral (PVB), tetrahydrofuran (THF) and the other reagents are commercially available and of analytical grade.

Preparation of TiOPc nanometer particles

TiOPc was dissolved in 98% concentrated sulfate acid first, then was added slowly into the water solution of a water-soluble polymer matrix with violent stirring at a temperature, during which the TiOPc fine particles embedded in the polymer were precipitated slowly. The dispersion system was washed by water until the pH ≈ 7, then precipitated, dried at 60°C in the vacuum, and the powder containing TiOPc nanometer particles embedded in a polymer was obtained.

Photoreceptor fabrication and photoconductivity measurement[6].

The dried TiOPc nanometer particles embedded in polymer were added into CH_2ClCH_2Cl solution dissolving α -NH and PC. The mixed suspension was coated onto a PA precoated aluminium plate with a thickness of about 30μm, and a function-separated single-layer photoreceptor(P/R) was fabricated that consisted of TiOPc nanometer particles as charge-generation material (CGM) and α -NH as charge-transportation material (CTM) (Fig.1a). The ratio of TiOPc, α -NH and PC was 1:20:20 by weight.

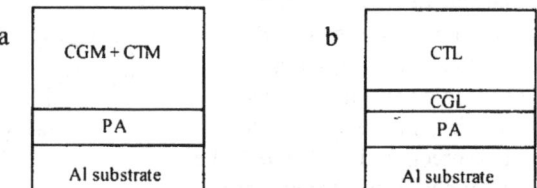

Figure 1   Structures of single-layered (a) and double-layered (b) photoreceptors

To a comparison, a function-separated double-layered P/R was made (Fig.1b), where the charge-generation layer (CGL) coating liquid is the TiOPc nanometer particles dispersed in PVB solution, and the charge-transportation layer (CTL) is coated with the CH_2ClCH_2Cl solution containing α -NH and PC (1:1, by wt.).

The xerographic properties of TiOPc nanometer particles are measured on a GDT-II model photoconductivity measuring device by the xerographic photodischarge method. Near infrared light of 760nm was applied as a light source, and the exposure intensity is 30μW. Two important parameters was obtained from the measurement: time from the initial potential to half under exposure ($t_{1/2}$) and half discharge energy ($E_{1/2}$). A good photoconductive material should have low $t_{1/2}$ and $E_{1/2}$ values, especially, the smaller the $E_{1/2}$ value, the higher the photosensitivity.

RESULTS AND DISCUSSION

TiOPc nanometer particles embedded in polymers, namely as casein, are obtained successfully by the method of LPDR. The optimal experimental conditions for preparing TiOPc nanometer particles were studied. The mean grain sizes of TiOPc particles and the particles morphology were characterized by AFM, TEM, UV/VIS absorption, and X-ray diffraction patterns. The results showed that the shape of TiOPc nanometer particles was spherical or oval-spherical, and the crystal state was amorphous. The red-shift in the absorption of TiOPc nanometer particles was observed in addition to the blue-shift which is the character of the inorganic nanoparticles.

The xerographic properties of TiOPc nanometer particles were investigated in function-separated single-layer photoreceptors that consist of TiOPc nanometer particles as CGM and α-NH as CTM. Table1 lists the photoconductivities of the P/Rs from various TiOPc nanometer particles. To single-layer P/Rs from TiOPc, we find that, TiOPc with average particle size of 50 and 20nm has shorter $t_{1/2}$ (0.35

Table 1   Relationship between particle size and photoconductivity of P/Rs from TiOPc

average particle size (nm)	P/R	$t_{1/2}$ (s)	$E_{1/2}$ (μJ/cm^2)
281	S-L	0.71	5.33
	D-L	1.07	8.09
50	S-L	0.35	2.63
	D-L	0.33	2.48
20	S-L	0.18	1.35
	D-L	0.25	1.88

note: 1) S-L and D-L mean single-layer and double-layer structure, respectively;
2) Thickness of a single-layer P/R is about 30μm; To a double-layer P/R, the thicknesses of CGL and CTL are about 0.5μm and 30μm respectively.

and 0.18s) and lower $E_{1/2}$ (2.63 and 1.35μJ/cm^2) values when compared to that with average particle size of 281nm (0.71s and 5.33μJ/cm^2), indicating much better photoconductivity. It is also noticed that the $t_{1/2}$ and $E_{1/2}$ values decrease with decreasing the size of TiOPc particles. About 2-fold increase in photosensitivity (from 2.63 to 1.35μJ/cm^2) is observed when the average particle size of TiOPc is reduced from 50 to 20nm. All these observations suggest that nanometer sized TiOPc owns excellent photoconductivity, and that the smaller the particle size of TiOPc, the better the photoconductivity. The same result can be drawn from the double-layer P/R of TiOPc as well (see Table1).

Photogeneration mechanism study showed that the photocarriers are generated at the CGM/CTM interface resulted from the CTM penetration into the CGL[7]. The larger the surface area of the CGM/CTM interface, the more the photocarriers generation sites, and consequently the higher the photosensitivity. There is no doubt that the surface area of the CGM/CTM interface increases with decreasing the size of TiOPc particles whether the P/R is single-layered or double-layered. So it is easy to understand that the photosensitivity is enhanced with decreasing the size of TiOPc particles to both single-layer P/R and double-layer P/R.

From Table 1, we also found that, when compared to the single-layer P/R (same size of TiOPc particles), the double-layer P/R has longer $t_{1/2}$ and higher $E_{1/2}$. It suggests that the photosensitivity of the single-layer P/R is higher than that of the double-layer P/R. The result can also be explained in the view of the surface area of the CGM/CTM interface. The surface area of the CGM/CTM interface in the single-layer structure is obviously larger than that in the double-layer structure, leading to better photoconductivity.

ACKNOWLEDGMENT:

This work was financed by the National Natural Science Foundation of China (No.59603003 and 69890230).

References:

1.   Wang,Y. and Herron,N., 1991, *Journal of Physical Chemistry*, **95**, p.525
2.   Koyman,T., Miyazaki,H. , and Anayama,H., 1990, U.S.Pat.4,895,782
3.   Zhang,X., Jenekhe,S.A., and Perlstein,J., 1996, *Chemical Materials*, **8**, p.1571
4.   Akamatsu,K. and Deki,S., 1997, *Nanostructured Materials*, 8(8), p.1121
5.   Moser,F.H. and Thomas,A.L., 1983, *The Phthalocyanines*, CRC, Boca Raton, FL
6.   Chen,H.Z., Wang,M., and Yang,S.L., 1997, *Journal of Polymer Science, A:Polymer Chemistry*, **35**, p.959
7.   Umeda,M., Niimi,T., and Hashimoto,M., 1990,   *Japanese Journal of Applied Physics*, **29**, p.2746

32 Dielectric Properties of Au/SiO$_2$ Nanostructured Composite Material

Tong B. Tang[1] and W. S. Li
Department of Physics, Hong Kong Baptist University, Waterloo Road, Kowloon, Hong Kong, China
[[1] tbtang@hkbu.edu.hk; http://www.hkbu.edu.hk/~matsci]

1.1 ABSTRACT

Au/SiO$_2$ 0–3 composite has been fabricated via a sol–gel process. Measurements show that above a certain level of Au doping, the dielectric constant K increases significantly while the dielectric loss remains relatively low. Moreover, for the same doping fraction, the smaller the size of Au clusters, the higher is K, as expected from the percolation theory. Au doping also modifies the temperature dependence of K such that it peaks near the room temperature, a fact of practical importance. The dielectric strength exceeds 10^5 Vcm^{-1} unless the doping fraction approaches the percolation threshold, and the optimum characteristics ($K \approx 10^2$ and tan$\sigma \approx 10^{-3}$) are offered by the composite with 14 vol% gold of 6 nm mean size.

1.2 INTRODUCTION

SiO$_2$ doped with gold clusters has been fabricated via a sol–gel process, and shows promise as a candidate material with gigantic dielectric constant. This 0–3 composite has previously been studied for its high third–order nonlinear optical response (Ricard *et al.* 1985; Hache *et al.* 1998; Bloemer *et al.* 1990; Hosoya and Suga 1997), and it has been fabricated via melting method (Kozuka and Sakka, 1993), ion implantation technique (Fukumi *et al.*, 1991), reactive sputtering (Maya and Paranthaman,1996), Granqvist–Buhrman technique (Grannan *et al.*, 1981), and a route using polmer precursor (Olsen and Kafafi, 1991). A sol–gel process for its synthesis has, however, the obvious advantages of good control of dopant concentration, homogeneity in composition, ease in production scale–up, high throughput, as well as low cost. This processing method has therefore been adoped in this work, which aims to identify its optimum performance.

1.3 EXPERIMENTAL DETAILS

Several series of samples were produced from solutions of $AuCl_3$, tetraethoxy-silane (TEOS) and C_2H_5OH in distilled water. The level of Au doping was controlled by the molar ratio of $AuCl_3$ to TEOS. After the hydrolysis of TEOS, a suitable amount of aqueous NH_4OH was added dropwise to accelerate the gelation. Then the solution was poured into plastic containers inside a semi–hermetic box, with ambient humidity gradually decreased to prevent cracking in the forming solid. The porous solid thus formed was stored in a dynamic vacuum followed by heat treatment at 200 °C for half an hour, to have its volatile and water residues removed and its $AuCl_3$ reduced, before being grounded into powder and then compacted under 1 GPa. The pellets so obtained were fired at 600 °C for 5–6 hours to achieve densification.

Gold electrodes were sputtered onto the samples. AC impedance, as functions of both frequency and temperature was measured with a Hewlett-Packard 4284A LCR meter equipped with low and high temperature test heads. The frequency span was from 20Hz to 1MHz, and temperature ranged from 20 to 350 K. Dielectric strength was determined at a ramp rate of 1 Vs^{-1}, starting at a few hundred volts below the anticipated breakdown voltage.

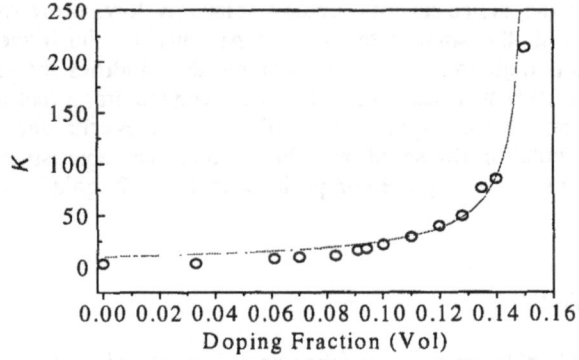

Figure 1.1 Relationship of dielectric constant with gold doping fraction: experimental data (open circles) and fitted scaling law (smooth curve)

1.4 RESULTS

Complex impedance measurements indicate a weak dependence of dielectric constant K on frequency in all samples, within the frequency range under study. The dependence on Au doping fraction of K at 1 kHz (as representative example) follows a scaling law (Figure 1.1):

$$K = C \, \varepsilon^{-s} \qquad (1.1)$$

where C is a materials parameter, $\varepsilon = (V_c - V) / V_c$ is the reduced volume fraction, V_c being the percolation threshold, and S is the critical exponent. Curve fitting leads to the values of 0.16 ± 0.01 and 0.86 ± 0.05, respectively, for V_c and S in our composite system. The V_c value matches the theoretical derivation (Grannan *et al.*, 1981) of 0.15 perfectly, but the S obtained is a bit larger than the theoretical estimate by Straley (1977), namely 0.70. This discrepancy can be attribued to cluster size distribution.

Then, at the same doping fraction (eg. 0.13), the smaller the mean size of Au clusters, the higher is the dielectric constant of the composite (Figure 1.2). The mean cluster size was estimated from X–ray diffraction peak broadenings at 2θ = 44.4° in samples further annealed at 950°C for different lengths of time.

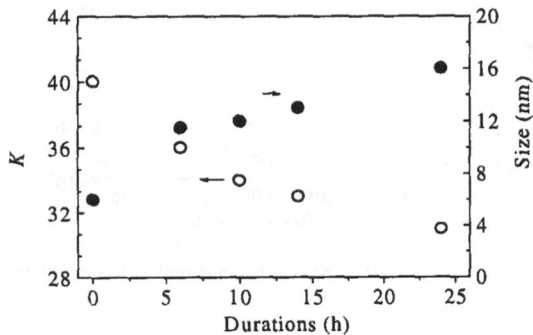

Figure 1.2 Variations of K (at 1KHz) and Au cluster size with durations of annealing at 950°C

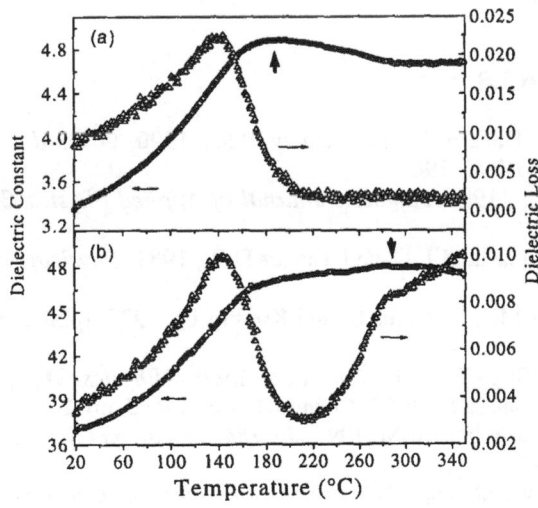

Figure 1.3 Temperature dependence of dielectric response at 1KHz, in SiO_2
(a) without and (b) with Au doping

Thirdly, results on the temperature dependence (Figure 1.3) show that the dielectric constant of SiO_2 has a maximum at 175K, but in Au/SiO_2 the maxmum shifts to near room temperature. Dielectric loss (Figure 1.4) in our materials system was generally low, except in samples with Au doping fraction of 0.15, which approaches the percolation threshold. There is a small local maximum near the fraction of 0.11, where a local minimum in dielectric strength also occurs. Apart from cases with doping fractions of 0.15 and beyond, the breakdown field F_{br} exceeds 10^5 Vcm^{-1}.

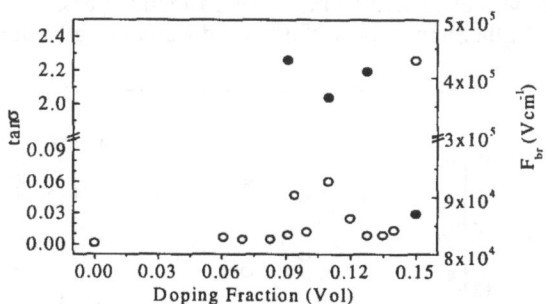

Figure 1.4 Dielectric loss at 1KHz (open circles) and dielectric strength F_{br} (solid circles)

1.5 CONCLUSION

As a dielectric material, SiO_2 containing 14% Au in mean cluster size of 6 nm provides an optimum performance of $K \approx 10^2$, tan$\sigma \approx 10^{-3}$ and $F_{br} \approx 10^5$ Vcm^{-1} at room temperature.

1.4 REFERENCES

Bloemer M.J., Haus J.W. and Ashley P.R., 1990, *Journal of American Optic Society* **B7**, pp. 790–796.

Fukumi K. et al., 1991, *Japanese Journal of Applied Physics*, **30**, 4B, pp. L742–L749.

Grannan D.M., Garland J. C. and Tanner D.B., 1981, *Physical Review Letters*,. **46**, pp. 375–378.

Hache F., Ricard D., Flytzanis C. and Kreibig U., 1998 *Applied Physics* **A 47**, pp. 347–354 .

Hosoya Y. and Suga T., 1997, *Journal of Applied Physics*, **81**, pp. 1475–1480.

Kozuka H. and Sakka S., 1993, *Chemical Materials* **5**, pp. 222–228.

Maya L. and Paranthaman M.,1996, *Journal of Vacuum Science and Technology*, B **14**, pp. 15–21.

Olsen W. and Kafafi H., 1991, *Journal of American Chemical Society*, **113**, pp. 7758–7760.

Ricard, D., Roussignol P. and Flytzanis C., 1985, *Optics Letter*, **10**, pp. 511–517.

Straley P.J.,1977, 1977, *Physical Review B*, **15**, pp. 5733–5739.

Author Index

Subject Index

Milton Keynes UK
Ingram Content Group UK Ltd.
UKHW020027071024
449327UK00032B/2963